Reason, Science
&
Revelation

MANUEL M. CARREIRA SJ.

JULIO A. GONZALO

Ciencia y Cultura
Madrid
2014

Reason, Science & Revelation

1st Edition June 2014

© Manuel M. Carreira
© Julio A. Gonzalo

© Asociación Española Ciencia y Cultura
c/ Pavía 4, 1º D. 28013 Madrid (España)
Fax: (34) 91-4978579
www.cienciaycultura.com
E-mail: julio.gonzalo@uam.es
E-mail : aecienciaycultura@gmail.com

ISBN: 978-84-942740-2-2
ISBN eBook: 978-84-942740-3-9

CONTENTS

Foreword

Reason may be defined as the ability man has to think logically, to draw conclusions. It is related to good sense, and to sanity, and opposed to craziness. Science, as distinct to the Humanities is the ability to think systematically, with proper order and method about eh behavior of material objects. Scientific knowledge is therefore the result of realistic observation and its principled evaluation. Revelation, which has theological connotations, is the disclosure, sometimes a striking disclosure, of previously unknown information. Obviously, its intrinsic value depends on the authority of the source of that revelation and on the way that revelation is made.

This book is made up of a few chapters on topics directly related to reason, science and revelation taken out at the last minute from our previous book *Intelligible Design* (World Scientific: Singapore, 2013), a book coauthored by a number of distinguished authors specialized in physics, engineering, biology, history, and others disciplines.

A 14 pp. long report by a Senior reviewer entitled *Intelligible Design or unintelligible gibberish?* made a devastating critique of most contributions in our book, ending with a comminatory sentence addressed to World Scientific: "I can only repeat that were you to publish this book (*Intelligible design*), you would do massive, perhaps fatal damage to your reputation as a scientific publisher..."

Fortunately, after various ups and downs, World Scientific decided to publish our book, but we, as Editors, decided to take out a few chapters and to take the extra chapters apart together with some relevant additional material and publish them under the title *Reason, Science and Revelation*.

In the meantime it was published *A universe coming from nothing* by L. M. Krauss (with an afterword by Richard Dawking), a book which appears to be written in direct opposition to our previous book, *Everything coming out of nothing vs a Finite, Open and Contingent Universe* (M. M. Carreira and J.A. Gonzalo).

The book by L.M. Krauss is interesting, but we find it superficial in so far as it awakens a hope that there will be an answer, something that is not clearly given. The very title is misleading: the NOTHING turns out to be a **physical vacuum full of energy** and subjected to concrete physical laws. The basic problem -the need for an initial determination of properties for something whose very nature does not imply them- is not even mentioned, but those properties must be determined by an external previous cause.

Modern Cosmology teaches that the age of the Universe is finite in the past, and that its present material content is also limited. By the most obvious logic, if there was a Beginning, we should ask for a Sufficient Reason why the Universe exists at all and for its actual properties: whatever does not exist by itself (as an eternal and immutable Universe would) must have an **extrinsic** cause from which existence depended, a cause that should also determine the concrete way existence occurs (the various components of the Universe and their way of acting).

Even pure scientists have admitted that whatever is **changeable** -as matter is- by its changes is proving that it isn't determined to exist in a unique way. Therefore it CAN and MUST be adjusted extrinsically to exist in a particular concrete way. Everything else that later occurs must be a consequence of that initial state that is **misleadingly** called "NOTHINGNESS".

It is still always true that "from NOTHING, nothing develops": nothing comes out, because there is no content to start with.

As none other than Voltaire pointed out, an atheist culture is not viable.

M. M. Carreira and J. A. Gonzalo

.

Start with a right triangle:

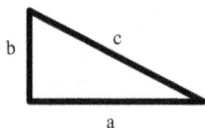

Area = c^2:

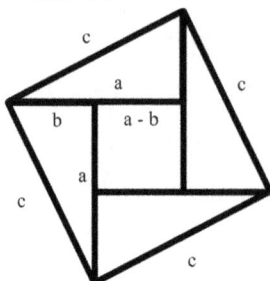

And arrange four of them:

Now move two of them:

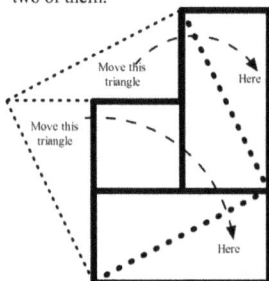

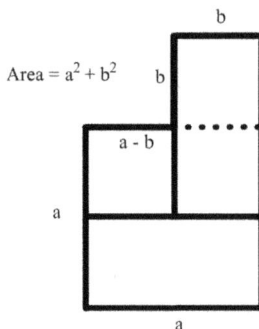

Area = $a^2 + b^2$

Early Proof of the Pythagorean Theorem

1. Proofs & Demonstrations
by Julio A. Gonzalo

What does it mean to "prove" a proposition? Clearly, to put it in terms of another proposition which is more *intelligible*, more understandable. The process can be repeated once, twice…many times. But an *indefinitely long* chain which ends with the last proposition hanging in the air is *not* satisfactory. And a closed chain of propositions in a circle would not be satisfactory either.

In order to say *something* conclusively, of course, it is necessary to start with an *affirmative proposition*, as evident as possible, which is accepted without proof. It is not always clear which proposition is the best as the starting point, and, consequently, there might be two or more *legitimate* starting points.

There are of course, many different *modes* of proof applicable in one or more of the various fields of knowledge. In particular: a) *deductive proofs* (used in logic and mathematics); b) *inductive proofs* (used often in the physical sciences); c) *empirical/historical proofs* (used in medicine, in the humanities, in sociology and in history); and *metaphysical proofs* (used in the realm of the ultimate questions: for example, why there is something and no nothing?).

In a *deductive proof* one reasons from the general to the specific. In an *inductive proof* the reasoning goes the other way: from particular facts to general conclusions. Inductive and deductive proofs properly formulated are undeniable, almost tantamount to tautological statements. They are common currency in mathematics and often so in physics or chemistry, but relatively rate in geology or biology, and practically absent in the historical, psychological and political sciences.

Empirical and historical proofs may not have the high degree of evidence provided by deductive and inductive proofs but they may still have a highly compelling character, A good detective which investigates a case of murder putting together the relevant facts and their timing excluding reasonably all possible alternatives but one may not arrive to absolute mathematical certainty may still be a perfectly valid proof.

A metaphysical proof belongs in the realm of metaphysics which is beyond the purely physical or sensory realm. Metaphysics is that part of philosophy which deals with reality as such, its fundamental properties and its first causes.

The metaphysical approach to essence and existence may be more abstract than the purely physical approach, but is not less real.

In adition to material realities there are realities there are realities, like intellect, will and freedom, which are non material but undeniable. When we define material reality we must express a suitable way of checking the behavior of matter as due to the four interactions which physics recognizes (gravitation, electromagnetism, strong and weak nuclear forces).

We can illustrate this with some *examples*:

Deductive proofs:
Geometrical: "The sum of the three angles of any triangle on a plane is equal to two right angles (180°)".

"*Reductio ad absurdum*": "There are infinite prime numbers".

"*Complete induction*": "All integers are prime or product of primes".

Algebric: "The two roots of a quadratic equation $(ax^2+bx+c=0)$ are given by $x=[-b \pm (b^2 - 4ac)^{1/2}]/2a$.

(Any good and sufficiently complete Encyclopaedia should give account of these proofs and of the other proofs below.)

Inductive proofs
Physical/Geological
Inertial motion: In the absence of friction or external force, a massive body remains at rest or keeps moving at constant speed in the same direction.

Gravity: A massive body on the earth's surface experiences a gravitational force given by $F = m (GM_T/R^2) = mg$, where m is the mass of the body and $g=980$ cm/sec^2 is the gravitational acceleration at the earth's surface, determined solely by G (Newton gravitational constant), M_T (the earth's mass) and R (the earth's radius).

Atmospheric pressure: The earth's atmosphere (made up mainly of molecular nitrogen and oxygen) exercises a pressure on the earth's surface equivalent to that of a mercury column of 760cm at sea level (under standard conditions).

6

Finite velocity of light: The light's velocity is finite. Therefore light traveling different interplanetary distances would take different times. Reasoning this way Roemer (1675) made a correct determination of light's velocity as $c \approx 3x10^{10}$ cm/sec with about 25% error.

Astronomical

Earth's radius: Parallel Sunlight rays incoming (1) nearer the equatorial earth's plane and (2) further from the equatorial plane make different angles with the local perpendicular to the earth's spherical surface at points (1) and (2). Measuring the respective angles at the two different locations in the same day, and knowing the distance between both, the earth's radius (r_E) can be obtained. Reasoning this way Eratosthenes (240 b.C) obtained, with surprising accuracy, the equivalent to $r_E \approx 6350$ km in Greek "stadia" (800 km $\approx$ 5000 "stadia").

Moon's radius: At a Moon eclipse, comparing the time needed by the Moon to disappear in the shadow produced by the Earth with the time needed to reappear again (emerging from the Earth's shadow) provides enough information to estimate the Moon's radius in comparison with the Earth's radius. Reasoning this way Aristarcus of Samos (c. 250 b.C.) estimated r_M (Moon) as slightly above one fourth of r_E (Earth).

Distance Earth – Sun: The triangle Earth (E), Moon (M), Sun (S) becomes rectangular when *exactly* half-moon is viewed from the earth. Therefore, if one were able to determine *precisely* that angle (α) at the earth's vertex (an angle very slightly less than 90°), knowing the Earth – Moon distance (d_{EM}), it would be possible to determine the Earth – Sun distance (d_{ES}). Aristarcus of Samos (c. 250 b.C.) attempted the determination of d_{ES} in this way but due to the low accuracy in the determination of α, could only conclude that d_{ES} was much larger than d_{EM}.

Empirical / historical proofs

Historical

Accurate translation of old Egyptian hieroglyphs: After Napoleon invaded Egypt, the discovery of the "Rosetta Stone" (1799) provided the "key" to the translation of Egyptian hieroglyphs. It contains a decree of King Ptolemy V in *Greek, demotic Egyptian* and *hieroglyphic script.* This opened the way to the translation of many other ancient documents. (See f.i. "The Lion Encyclopedia of the Bible", Lion Publishing Corporation: Batavia, Illinois, New revised edition, Reprinted 1987. A *color photograph* of the *Rosetta Stone* is given in p.34)

Historicity of the Gospels: The earliest known fragment of the New Testament in a papyrus, copied around 130 AD, which contains part of the Chapter 18th of John's Gospel. Records of contemporary census fit very well with Luke's account of Jesus birth in Bethlehem. The earliest references to Chrestus (Christ) and Christians by pagan writers are in Suetonius "Lives of the Caesars", AD 121, and Tacitus (AD 51-117). "Annals", 15.44. Copies of famous Greek books such as "The Iliad" and "The Odyssey", have been found among the papyri contemporary with John's Gospel. (See f.i. "The Lion Encyclopedia of the Bible", Ibidem, pp 62-63).

Christian medieval roots of modern science: Jean Buridan, in his Commentaries to Aristotle's: *"On the Heavens"*, written around 1350 AD, formulates for the first time *the law of innertial motion*, Buridan's work as well as that of his disciple Nicole Oresme, became well known throughout medieval Christendom. In the ensuing years Copernicus, Galileo and Descartes (contemporary of Newton) provided the subsequent steps for the formulation of Newton's *"Principia"*, which gave birth to modern science (See f.i. S.L. Jaki, "Bible and Science", Christendom Press: Front Royal, 1996, p. 103 ff.) as shown by the great French theoretical physicist and historian of Statics and Dynamics Pierre Duhem.

Metaphysical proofs

Proofs of statements such as: a statement is either *true* or *false*: there is no *medium* term.

To affirm positively that there is no truth (*relativism*) is self-contradictory.

Human intellect, which is capable to go beyond sensations, images, and perceptions to *ideas* (*concepts*), *judgments* and *chains of reasoning (syllogisms)* goes well beyond the capabilities of animals and "robots".

God's existance

The arguments to prove *God's existence* and some of his attributes are *metaphysical* in character. His essence is obviously *beyond* our intellectual capabilities. But by way of analogy, from the *created world*, and from our own inner world (our *conscience*) as Saint Augustine pointed out ("noli foras ire, in interiore hominis habitat veritas" – "do not go outside, in man's interior dwells truth"), men can know *God's existence* and some of his *attributes*.

There exists *something*.

This something consists in a world exterior to me and my own interior self.

The *external world* is changing, contingent and orderly. Therefore, being *contingent* (not necessary) it cannot have given itself its existence. This *implies* an immutable necessary *Creator*, capable of *manteining* continuously the actual existence of that external world. This is the Cosmological argument developed along classic Aristotelian arguments by St. Thomas Aquinas in his famous five ways to God.

The *interior world* (i.e. the *subject*) is conscious of his own *continuous* existence, an existence which takes place in a higher plane, and implies intelligence, will and freedom, as well as capability to perceive truth or falsehood, good and evil, dignity and indignity. Therefore the subject, even being obviously capable of falsehood or error, of wickedness and indignity (objectively considered) cannot be the one who gives himself *that standard* producing in him "*good*" or "*bad*" conscience. This *implies* a Creator, the *origin and the source* of that conscience capable of *maintaining* the continued *awareness* in the subject of his own capacity to perceive that truth, good and dignity, in himself and in the exterior world.

That Creator of the external world and of the personal subject, responsible of the *possibility* and *continuous reality*, coherence and harmonious compatibility of *external world* and *subject* is what everybody calls God.

Remarks:

The possibility of the real existence of *free subjects* is not easy to make compatible with the determinism and the apparent chaos pervading nature. Both, determinism in the natural world and free will in the subject, are undeniable *primary realities*. And man's free activity is clearly a pointer to the creative freedom of God.

If I look for a sufficient reason for creation I must find it in the creation of free, personal beings capable of relating freely with the Creator.

The arguments to prove God's existence have a long and distinguished tradition, involving throughout the centuries such a first rate minds as those of Plato and Aristotle, St. Augustine, St. Anselm, St. Thomas Aquinas, St. Bonaventura, Descartes, Suarez, Leibnitz and many others, including Gödel.

Gödel based his proof upon Leibnitz's "A priori" proof. Therefore, being perfectly consistent from the logical point of view it is ineffective as a proof of existence, in as much as it makes a jump from the conceptual order to the order of actual existence.

Gotfried W. Leibniz (1646-1716) arguably one of the two greatest geniuses of the "century of genius" (the other is Isaac Newton) proposed a reformulation of the old and famous "ontological argument" of St. Anselm, which can be put succinctly as follows:

"God is the most perfect being which can be imagined / But the most perfect being must posses the perfection of existence in addition to all other perfections / Therefore God exists."

The implicit paralogism is subtle but fairly evident: in the purely *conceptual* order the conclusion is *true*, but the jump to the *existencial* one is *not valid*.

Gödel's argument, based upon Leibnitz's reformulation of this argument is much more elaborate but, as long as it is an "a priori" argument is not valid either. Only real existing beings, not merely conceptual entities, are valid starting points to demonstrate the existence of anything.

It may be pointed out that Leibnitz`s other proofs, based, respectively upon the reality of eternal truths, the contingency of ordinary beings and the pre-stablished harmony in the world around us, are not "a priori" proofs, and therefore valid in principle, to infer an actually existing omnipotent Creator.

Isaac Newton (1624-1727) seems to be in agreement with the cosmological argument[1] when he says in his "Principia"
"This most beautiful system of Sun, Planets and commets could only proceed from the counsel and dominion of an intelligent and powerful Being".

The 20th century astronomical discoveries are, of course, compatible with a created universe. The standard "big-bang" theory is compatible with a created universe. The "steady state" theory (its rival for a quarter of a century) assumes a universe eternally expanding at constant density, and was not compatible with a universe created in time. It should be pointed out that the picture emerging from these scientific discoveries, had nothing to do with a naive "concordism". The language of Genesis I was addressed to the ordinary Hebrews living two thousand years ago, and it is not meant to be a scientifically accurate statement for today's standards. Which does not detract from the validity of its central message: God is Lord of *all*, in *heaven* and on *earth*.

REFERENCES

[1] "Principia", translated by F. Cajori, U. of California Press (Berkeley, 1934) p. 544

[2] See f.i. Fred Heeren, "Show Me God: What The Message From Space is Telling us About God" (Searchlight Publications: Wheeling, Illinois, 1995).

2. Finality in Science & human life

by Manuel M. Carreira, SJ.

C we are dealing with terms and concepts that are used in multiple contexts and in different fields, it becomes necessary to analyze in detail their meaning when applied to Theology, the interpretation of scriptural passages, and as part of our ordinary language. Otherwise it is easy to fall into presuppositions that lead to wrong ideas regarding the relationship between Science and Faith, and the role of miracles within Christian Theology.

Different meanings of the term "Faith"

a) First of all, FAITH is *a way of obtaining knowledge*, not through direct personal experience or reasoning, but through *the testimony of others* who are considered as possessing some truth that can be shared without distortions due to prejudice or the desire to mislead. Trustworthy witnesses are the basis upon which innocence or culpability is decided in a judicial court; even those who are called as experts to examine material evidence must be trustworthy when they present their testimony.

This faith does not have an automatic relationship with the field of Religion: it is of much more general applicability, as *the universal means to acquire culture* from different sources, of other human groups and of other ages. Without this human faith the knowledge of history would be impossible, as it would be to find out about distant places, and to acquire the contents of all the specialized sciences where personal effort cannot give an answer (e.g., Atomic Physics, Medicine, foreign languages...practically everything we know)

Human faith is a basis for *certainty* even *against the data of my own sense experience or the spontaneous arguments of "common sense"*. We are sure that ordinary matter, even of our own body, is a cloud of infinitesimal particles in fast motion within an almost empty space; we also accept that the energy of an impact between particles can be converted into new particles. The consensus of experts who have no reason to deceive is a *necessary and sufficient condition* for scientific progress in any field, allowing us to build upon the contributions of past generations.

Certainty based on human faith is possible even when our assent is given to something *that we do not understand*. "Nobody understands Quantum Mechanics" (Richard Feynman) or the fact that General Relativity is incompatible with it, while both are pillars of Modern Physics with multiple experiments that prove their validity in their own fields. Such is the power of conviction when there are reliable sources for our faith.

Within the area of Theology, the basis for our acceptance of Christianity is the *historical* fact of Christ's presence and teaching in Palestine 2000 years ago. This fact must be established by the same methodology and with the same criteria that we apply in order to know about Julius Caesar or Christopher Columbus. Christ's teachings must be established by the same steps that we require to find out the teachings of

Socrates: *the disciples who lived with the Master give us their testimony about his actions and words.* The disciples of Christ were witnesses who preferred death to any denial or compromise when transmitting their own experience.

This human, historical, faith is the *rational* foundation of Christianity; without it, the acceptance of its message would be absurd. We must have *objectively valid proofs*, otherwise our faith would be only some kind of vague, emotional, private reaction without universal applicability (as the Pope teaches in his Encyclical *"Faith and Reason"*) or simply a mythology of a poetic nature that has no bearing upon real life.

A most important part of the historical testimony about Christ is the insistence of the witnesses upon facts that proved *God's sanctioning* of the activities, teaching and personality of Christ. Those facts -*miracles*- were *required* in order that accepting his claims would not be considered irrational. Whoever says that he is the Son of God, equal to God, existing before Abraham, of higher dignity than the prophets, entitled to forgive sins, *must provide undeniable proofs of those claims*. Otherwise, it would be sheer madness to just take him at his word. Faith, in Catholic Theology, is called *"rationabile obsequium"*, a *rational response* to truth.

Once the divinity of Christ is accepted, his teachings have the highest degree of certainty, giving rise to *divine faith* which will accept even things that surpass our understanding. If we cannot understand even the behavior of matter in our laboratories, or our own body, we should not be surprised when God's nature and plans seem impossible to fathom. The Creator of the Universe is not just a "superman" understood as a merely human enlargement of our own persons, but something totally beyond our powers of imagining or extrapolating our experience.

The testimony of the apostles and disciples reaches us after twenty centuries, instructing us about Christ and his teachings. If this is God's plan for our finding the way to Him, *it is necessary that the message be unchanged*, so that no errors might be introduced by using other languages, or by personal interpretations, the loss of written texts, commentaries or arbitrary additions. This *requires a teaching ministry*, with authority due to God's own appointment and assistance, to provide a guarantee of absolute and total fidelity to the original message of Christ. Any group that presents God's word as subject to contradictory interpretations is, by this very fact, admitting that its teaching is of merely human origin and value: God cannot contradict himself.

Christ promised to his Apostles the Spirit of Truth, to keep them from falling into error or into a human "wisdom" that would add or detract from his teaching. That is why he could say *"whoever listens to you, listens to Me, whoever despises you, despises Me"*. Only the Catholic Church claims to have this guarantee of fidelity to Christ's message, not because of any self-assurance of being wiser than others, but because of the promise of the Spirit. This is why the encyclical letter *"The Splendor of Truth"* is addressed to the Bishops –successors of the Apostles- and not to the theologians: the Church can entrust theologians with explaining the faith, but they do not have a promise from Christ to keep them from falling into error. If they depart from the teaching tradition of the Church, they have no right to present themselves as *catholic* theologians, no matter how much they might know about the critical study of texts or the cultures of ancient times.

Since God is infinite Wisdom and Truth we may rest assured that the Christian message will never contradict scientific truths *in any field of human knowledge*. Apparent conflicts have been the result of taking as science or dogma personal interpretations that go beyond scientific data or the teaching of the Church.

b) The second meaning of FAITH rests upon the first, but instead of pointing to a means of obtaining new knowledge, it focuses upon the effect of that knowledge *on our behavior*. It can be understood as a trust or confidence that affects our *free will* (thus implying responsibility, for merit or guilt) and our heart. In our daily way of speaking, I might say *"I have a lot of faith in this medical doctor"* not because I expect to get new knowledge of Biology or Anatomy, but because my conviction of his expertise and honesty gives me the assurance of being helped in my sickness. This same use of the word "faith" can be applied to the expected benefits of a fitness program, a particular drug, a traditional remedy.

St. Paul uses the word FAITH in this sense when he explicitly writes *"Faith is the hope for the eternal goods"*. This *faith-confidence* presupposes the first step (faith due to testimony) and it may entail a *re-structuring of our personal life* to follow the teachings and ideals of a leader, be it in politics, philosophy or religion. This behavior will be *rational* only if the person to whom such trust is given *has proven that it is deserved*, both because of an outstanding merit in a particular field and because of the personal honesty and truthfulness of the leader toward those who want to give their allegiance to the admirable ideals that are presented.

This was the case with those who followed Christ, totally confiding in Him and *freely and rationally* deciding to adjust their lives to his teachings, because they were norms and truths given by God. The divine faith, embracing a Revelation received from Christ, rests upon the infinite trustworthiness of God himself. And those teachings do not remain in an abstract level of classroom learning, either of history or even of theological ideas, but rather become *personal life and blood* because there are clear credentials showing that such way of life is God's will: credentials that *prove Christ's unique dignity with miracles, especially his own Resurrection*. This was the reason for the behavior of the Apostles and of all those who during centuries have followed Christ to the highest levels of sanctity and even martyrdom, *living the logical consequences of their faith in the Lord*, known and loved and made the only worthwhile norm of life, by a personal and fully responsible decision.

This is *"the Faith that can move mountains"* because its strength is God's Omnipotence. The Lord never refuses his help to those who have complete confidence in Him, especially when God moves them to ask for a particular favor at some point, with the certainty that it is His will to give what is requested. The swift growth of the early Church, even amidst the persecutions from Jews and pagans, cannot be explained without accepting the impact of miraculous events that the *"Acts of the Apostles"* describes after Pentecost. We should not forget that the followers of Christ, entrusted with the mission to go to the whole world to proclaim his message, were a few men without any academic credentials or social standing, weak fishermen or country folk for the most part.

But we should also keep in mind that through the history of the Church many saints, who greatly influenced its development - St. Dominic, St. Francis Assisi, St. Ignatius, St. Therese of Avila- never performed a miracle in their lifetimes. This does not imply a lack of faith: the gifts of the Spirit are meant for the good of the Church, not for the personal glory of any member. It is possible –theologically- that a miracle might occur even through an unworthy minister, because its purpose is not to prove the holiness of the instrument, but the mercy and power of God.

c) The third meaning of FAITH concerns, not a human response, but a *gift* from God, a *Theological Virtue*, unattainable through human effort and without any visible effects either on the mind or the will of the recipient. It is given in Baptism, even to a baby who is unaware of the sacrament and does not acquire new knowledge or have any confidence as a result of this gift. It belongs to the supernatural realm: it is, in a certain sense, a *graft of divinity* that raises to the level of eternity every act of the person, incorporated to Christ to become a member of his Mystical Body, a child of God. Through this Faith one becomes a member of the Church in a permanent way, stamped forever with a seal that indicates God's choice, in such a way that even the most hardened sinner still receives graces, moving the will to repentance and leading to reconciliation with God's people. Only the sinner who apostatizes renouncing the faith loses this vital connection in a kind of spiritual suicide, but -even then- Baptism cannot be performed again if the sinner repents.

Because this *theological virtue of Faith* requires knowledge and free acceptance in order to be fruitful, the Church does not allow Baptism for a person who lacks instruction in the contents of Christian teaching, unless there is a sufficient reason to expect that the instruction will be obtained: this is applicable to children whose parents and godparents explicitly guarantee their Christian upbringing after Baptism. *We are not dealing with some kind of magical rite*, but with a joint effort in which human activity is the means through which God bestows his graces, even if all our efforts are infinitely inadequate for the purpose of reaching Him.

This is the Faith that is defined as *a gratuitous gift from God*, independently of any human effort or merit, but a gift that God *never refuses* to those who have done what they could to find Him. In the normal way of acting of divine Providence – with respect for our freedom and the activity of the created world- this Faith will be obtained after the two previous stages, with a sufficient knowledge of the life and teachings of Christ and the confidence that He is the only Way that God wants us to follow to find Him. Being a *"Virtue"*-which means *a force, a power, an active principle*- it is not something in the realm of either sense or rational knowledge, but rather a new power *to act at a divine level*. But activity follows nature: this is why we are taught in Theology that by Baptism we become *sharers in the divine nature*.

Neither angels nor any created being can partake of God's nature except through the generous gift of the infinite source of Goodness and Life. This is *saving Faith*, that introduces us into the way that God lives, and that cannot exercise its power except in union with Hope and Charity, the Love of total self-giving which is the very life of God and our eternal happiness.

Since this is the eternal destiny of all human beings of any place and time, we can be sure that God will give enough light and grace to everybody in order to have a truly responsible act of accepting God's will, even if personal or cultural constraints seem to make that knowledge and response practically impossible. We do not know how God works upon the human soul, perhaps at the moment of death, but we can say that His infinite justice and love truly imply a real desire for the salvation of everybody, a salvation that includes the incorporation into the Mystical Body of Christ.

Science and miracles

Science, in the modern technical use of the word (as different from the Humanities) is the *study of the behavior of matter —its interactions- through observation, experimentation and measurement.* This study discovers ways of acting that are more or less common and that occur with such regularity and constancy that *they must be due to what matter IS,* at diverse levels of its structure. Statements expressing in a generalized form those ways of acting are called "Laws of Nature", which are not some norms imposed from outside, but the necessary consequence of the very nature of material things.

Because our measurements can never be totally exact and our ability to observe nature is limited, laws are always stated in a restricted form, indicating margins of error and the levels of structure where the law has been verified. As an example, the Law of Universal Gravitation cannot be verified within an atom, and whatever might occur within a Black Hole will always remain undetected. But with those limitations - due not only to our technology, but to nature itself and to scientific methodology- we can assert that those laws are always obeyed in each particular instance. When the knowledge of all necessary parameters is impossible, it becomes acceptable to speak only of *chaos* or *chance*, even if neither term describes a physical force or gives *a reason* for the observed behavior. They rather refer to instances where the detailed prediction is impossible, either because we cannot obtain data at the atomic level or because in very complex systems the mutual dependence of multiple components renders their evolution unpredictable in the long run.

The logical consequences of studying a Universe that appears finite, both in spatial dimensions and in its past temporal evolution, oblige science to admit a *beginning from nothing of a material order*. There was no "previous stage" that would determine initial parameters or laws: only the word "creation" correctly indicates the infinite step from *nothing to something*, and creation requires the total determination of the created being by its Creator. Creation is a free act, not imposed by any type of self-development of a Creator who cannot be immersed in time and change, and who is not a part of the material world. Created beings, not having in themselves the reason for existence, need to be kept from returning to nothingness by the *conservation* which implies a *constant dependence* upon the source of their being.

While created matter acts according to its properties, and the Creator will not change things arbitrarily, it is always possible to have an extraordinary event for a sufficient reason. To affirm that matter must always act in a certain and fixed way, *even independently of its Creator*, implies a philosophical absurdity. If our own free activity can change the way things occur (for instance, when we apply a flame to an object that could exist indefinitely without burning), it defies all logic to say that the Creator cannot have the right to act upon things that he created and that he maintains in existence. Science is not rendered impossible because we admit the possibility of our own free activity; we can say the same, with a better reason, about the activity of the Creator, that cannot be limited by our norms.

It is also quite wrong to say —as a scientific statement- that in the world of material activity *anything can happen* if we compute the probability of an outcome during a sufficiently long time or in a large number of cases. This implies the denial of any true relationship between cause and effect, thus destroying the foundations of physical science, which always seeks *a reason for the order and constancy* of behavior that is observed. Chaos and chance are never a reason for regularity, and their application to complex systems still *has to save the conservation laws* that make science possible: conservation of mass-energy, of linear and angular momentum, of net electrical charge. To say otherwise renders science meaningless, ending frequently with an arbitrary appeal to a "need" to have multiple universes where every possible outcome of mathematical computations has to occur ...but cannot be verified!

Summing this up: Science is possible only because there is a fixed pattern of material activity, based upon the very nature of material things. But their existence and activity depends at every moment upon the Creator, who freely gave existence to matter and endowed it with those properties that determine its development. *For a sufficient reason of a higher order*, the Creator can act in extraordinary ways to change the normal pattern of material behavior, thus giving an indication of his powerful presence. This is what we accept as the *possibility* of a miracle. To deny *a priori* such possibility is arbitrary and without logical foundation.

The theological meaning of miracles

The *Catechism of the Catholic Church*, in its no. 156, underlines the tight relationship between *proofs* of the fact of Revelation, its doctrinal contents and the historical presence of Christ in our world, and the consequent acceptance of Revelation and its teachings when we embrace the faith. If God has made us in his image, *due to our rationality and free will*, he cannot ask that we abdicate reason when we encounter his historical manifestations, meant precisely to help us at those levels where reason alone is insufficient.

"In order that our acceptance of the faith might occur in harmony with reason, God wanted that *external proofs of his Revelation* be joined to the internal help of the Holy Spirit. Thus the MIRACLES of Christ and the saints, prophecies, the growth of the Church and its holiness, life and stability, are the most certain signs of divine Revelation, suited to our intelligence; they are reasons to believe, that show that *assent to the faith is not, in any way, a blind impulse of the mind*' (*Dei Filius*

DS 3008-9; cf. *Mc* 16, 20, *Heb* 2, 4)

Faith, in so far as it determines a person to follow Christ's teachings, has to be a *free act of the will*, never imposed from outside by any kind of authority: *"nobody may be forced to embrace the faith against his will"* (*Catechism* no. 160). But this is misinterpreted if it is taken to mean that faith *cannot be supported by rational evidence*: it is clear that the text quoted refers to *external human pressures*, social or political. In the internal forum, *prejudice* can lead to the rejection of proofs that are quite convincing and that show that a manifestation of God has taken place. Nor is this kind of "freedom" restricted to religious belief: in Nazi Germany the Theory of Relativity was considered wrong and unacceptable because of a racial prejudice (it was "Jewish science") and in Soviet Russia geneticists were forced to deny the stability of heredity because this tenet was incompatible with the communist dogma that social and political beliefs should be inherited.

That faith is *perfectly compatible with rational proofs* is obvious from the very first statement of the Creed: *"I believe in God...Creator of Heaven and Earth"*, a truth that the first Vatican Council defined could be obtained *with certainty* by the use of human reason. This logical certainty does not render faith impossible or empty. And the faith of St. Thomas, *touching the wounds of the risen Christ*, is true faith, even if he is rebuked for not accepting the sufficient testimony of other witnesses. In fact, rational proofs are *required* for our faith, and this is the purpose of the miracles that Christ presented as his credentials as God's envoy, the Messiah, the Son of God.

What constitutes a miracle?

The concept of an *apologetic miracle*, as accepted in Theology, refers to a *fact* that is objectively observable and that indicates God's direct activity upon nature. It is possible, at least as a hypothesis, to find phenomena that are externally verifiable with total independence of any prejudice or cultural conditioning and that cannot be explained within the framework of the laws that direct the behavior of matter. Not only because of our imperfect knowledge of nature, but because in the effect produced or in the way to attain it (for instance, *by ordering it to happen*) *all the well established norms of material activity –according to its forces and properties- are transcended.* In such a case -which will happen only within a religious context- we must accept God's activity *as the only sufficient reason* for the result: a *miracle.*

Since we require that the fact be objectively observable by anybody, we deal with *external* phenomena, not something merely subjective or transitory or depending upon the faith of the observer. To have a mystical experience, a vision of Mary, does not constitute an apologetic miracle, since nobody else can verify that it is happening (but there might be other accompanying phenomena that might be observable and that defy natural explanation). Possible effects of the human psyche upon the body are also difficult to distinguish from miracles unless there is a *sudden and permanent* organic change, mostly when dealing with persons who are highly excitable or prone to suggestions.

In this strict use of the term *miracle (apologetic,* leading to Faith) we cannot apply it, for example, to the Eucharist. There is nothing *observable* to indicate a change in the bread and wine before and after the words pronounced by the priest during the Mass: its truth is accepted by faith, not by physical evidence. On the other hand, there are serious reports of several instances when, during the Mass, the bread was visibly and permanently changed into flesh, observable even under the microscope, or the presence of blood was obvious to any observer. In those cases the visible changes would constitute a miracle.

As a different example, the "miracle of the Sun" reported from Fatima cannot be established as an objective change in the Sun itself: instruments observing it detected nothing special, and even among the crowd at Fatima not everybody could see the strange gyrations and light patterns reported by others. One could say that the fact that so many people saw the display cannot be explained by any natural cause, but the display could not be proven as a miracle.

The appearance of wounds corresponding to those inflicted during the crucifixion of Christ (the *Stigmata*) can be due to psychosomatic effects known in medical practice; the positioning of such wounds does not prove their placement in the body of the crucified Christ, but rather reflects the images -crucifixes- normally observed by the person exhibiting the stigmata. Miracles occur to show divine activity for the good of the Church, not to supply information that we should obtain by our effort.

As a perfect example of an apologetic miracle in relatively modern times, we can mention the "Miracle of Calanda" (1640), involving *the sudden recovery of a leg amputated below the knee two years and five months earlier.* Hundreds of witnesses, including the well known surgeons who performed the amputation, testified to the undeniable fact that the young man -Juan Pellicer- had suffered the accident that led to the loss of his leg, was seen for over two years begging with his crutches, and then recovered the leg while asleep at home. All of Europe marveled at the report, and even the King of Spain verified the fact when Juan went to Madrid at his request. But *prejudice* cannot deal with facts: the philosopher Jung acknowledges that the miracle has the best historical credentials that one could require, but because from its acceptance one would logically have to recognize the activity of God, it must be denied!

Miracles, in principle, could involve divine activity – recognizable as such- upon inanimate matter: the sea, the atmosphere, artificial or natural objects. Or upon human beings, in their external behavior, their bodies (healings) or their activities (even of the intellectual order: prophecy, knowledge of occult facts, of other languages). But in every case it is required that checks "before and after" establish the real facts, and then that *the result or the way to obtain it* be totally beyond the scope of material or human activity.

The miracles of Christ

The claims reported in the Gospels as made by Christ during his three years of public life are totally unique, not only in the entire Bible, but in the history of all religions as well. In a brief summary we can mention that he asserted repeatedly that:

- He is *"the One who is to come"*, the Messiah (*Mt* 11, 2-6; *Jn* 4, 26; 10, 24-25; *Lc* 7, 20-23)

- He has greater dignity and authority than prophets and patriarchs, Jacob, Abraham, Moses.

- He is *"The Light of the world"*, imparting eternal life (*Jn* 8, 12; 6, 51-58)

- He is *the only Way to God*, requiring a clear response pro or con (*Lc* 11, 23)

- He has the *divine* right to forgive sins (*Lc* 5, 24) and to work on the Sabbath (*Jn* 5, 19-30)

- His claims are superior to any family ties (*Mt* 10, 37)

- Only he knows God in an intimate way (*Jn* 17, 3-5)

- God alone knows him in his true personality (*Lc* 10, 22)

- He is the Son of God in a true and unique way (*Jn* 3, 16-18; 5, 18; 11, 30)

Those claims would be ridiculous if he didn't offer any *proof* of what he said, and to accept them at his word would have been irrational. But he appeals to his miracles as a guarantee of his mission:

- *"Go and tell John what you have seen…"* (*Mt* 11, 2-6)

- *"You still do not understand?"* is his rebuke to the disciples who fail to see the true meaning of the multiplication of the loaves (*Mt* 16, 8-11; *Mc* 8, 17-21)

- *"That you might realize that the Son of Man has the authority to forgive sins… get up and walk! "* (*Mt* 9, 4-6; *Lc* 5, 20-24)

- *"If I had not done things that nobody ever did, you would have no sin". "If you were blind, you would not be guilty of sin…but you see and don't believe: your sin remains"* (*Jn* 9, 41; 15, 24)

To deny the obvious meaning of all the texts in the Gospels concerning miracles, in the name of an arbitrary "demythologizing" that starts with the prejudice of some Protestant exegetes that *nothing of a supernatural order can be accepted,* is totally opposed to the relationship between faith and reason, besides lacking any scientific basis, because there is nothing to support it in the Gospels or in other believable sources. The purpose of the Gospels, explicitly stated, is to provide *reasons* to embrace the faith, through *the evidence of well attested historical facts (Lc* 1, 1-4). Only a starting point that reduces faith to a mere question of subjective feelings (many times rejected by the Church) can lead to affirm that we have to believe *without proofs;* on the contrary, *we need proofs.* This is not the same as saying that we believe only if we totally *understand* what we believe: *the proof does not deal with the understanding of the contents of the faith, but with the authority that makes it believable.*

The public activity of Christ, abundantly presented in detail in the Gospels, is centered on *signs* that are *visible facts that have a meaning* sought by Christ and quite clear for those who see things *without prejudice.* Such signs point to the power of God which guarantees the mission, the teachings, the holiness and the Person of Christ: *"The works that I perform, those testify in my favor" (Jn* 5, 36). They are miracles (*miraculum,* some admirable and marvelous event) that can be objectively seen *—they are nor something subjective-* and that have real effects, that are permanent at the normal level of permanence (a miraculous raising of the dead does not bestow immortality) and that cannot be attributed to the ways matter acts *according to its laws.* This last condition is the reason why they cannot be reproduced at will in an experiment: they only happen *by the free decision of God and in a setting that shows his activity for some supernatural purpose.*

Christ's miracles cover many different levels of activity. He knows the future, predicting Peter's denial, the cowardice of the Apostles who will leave him alone, the Passion in full detail, his own rising from the dead (*a prophecy that was clearly understood as such by his enemies*). He was also aware of hidden thoughts and of private experiences (Nathanael).

He showed his total control over inanimate matter: changing water into wine, feeding thousands with a few loaves, suddenly stopping a violent storm, walking on water. In each case the effects are due *only* to an effortless command and the Gospel remarks on the amazed reaction of the witnesses, and the comments of those who were experts on the subject.

He had an absolute power over health and life itself, shown in the sudden healing of multiple sicknesses, even at a distance (thus without any possibility of "suggestion" of the patient) and with added details that include the granting of sight to a man born blind, *together with the ability to interpret visual stimuli* (something that does not occur in the few cases when medical efforts give sight to the blind from birth). Faced with the miracle, the only answer of the prejudiced Pharisees is to disqualify and mock the one healed by Christ, but without being able to deny the fact or *to avoid his logical argument* about the holiness and mission of Christ.

In several instances he showed his control of life and death, returning to life even Lazarus, already decaying in the tomb. But his own resurrection is the key miracle, repeatedly forecast and later presented by the Apostles *as the main argument* for their faith. It was asserted as a *historical fact* in the full sense of the word, not because it happened before a notary public, but because it is the *necessary and certain inference* of the experience of witnesses deserving full trust: they saw Christ, dead and buried, and three days later *they saw him alive, they touched him and they ate with him.* This is the same as the historical reasoning for accepting that somebody was born even if there are no witnesses of the birth, or to say that somebody died even if the death itself had no witnesses: there is no doubt about those facts. *If nobody doubts death when seeing the corpse, there is no reason to doubt the resurrection when the same person is later seen alive, even if we don't understand how it happened.*

The evidence for the Resurrection of Christ was indirectly reinforced by the desire of his adversaries to prevent any fraud. The guards at the sepulcher made it impossible for anybody to go in or to take anything out. And confronted with the preaching of the resurrection by the Apostles, *nobody could show them to be liars by producing the corpse of Christ or some believable witnesses of its theft.*

This is why the Encyclical Letter *"The Splendor of Truth"* insists upon basing our faith on the *historical character of the Resurrection.* In the words of St Paul, *"if Christ did not rise from the dead, our faith is empty and we are the most miserable human beings" (1 Cor* 15, 16-19). The Apostles define their role as being *"witnesses to the Resurrection"* and before the Sanhedrin they claim to be obeying God's injunction when they preach it *(Act* 1, 22). Because of their preaching they are punished and finally put to death.

Only with the Resurrection of Christ can the Cross be seen as a triumph, overcoming its shame and the crumbling of the faith of the disciples. They had to be "beaten on the head" by *the evidence of their senses, even touching Christ and eating with him*: especially for a Jew, it made no sense to think of a metaphorical return to life or of a "resurrection" with a non-material body, which is also scientifically absurd, since *a body is only a material structure.*

Because of the Resurrection and of the previous miracles performed by Christ, our faith has a *rational* basis. *No further miracles are needed*: there is enough evidence for those who do not close their eyes to it. But we also gratefully accept God's continuing action in the Church to keep alive our awareness of the supernatural realities, which show God's love -even through marvelous means- towards those who believe and trust in Him completely.

NOTE

The concept of FAITH, *within typical Protestant theology*, has very different connotations from those presented in the first part of this paper. In a very concise description, Protestant doctrine holds:

That the human mind, due to original sin, is now *incapable* of attaining Truth, at least in the religious level.

That Faith is an assent given *without reasons to justify it*, or simply an act of the will *without a rational basis*. More drastically, it is said to be a *feeling*.

That the only source of Faith is the Bible, a book that *must contain in itself* its own proof of being revealed by God.

That this book allows different private interpretations, all of them equally acceptable, *even when they are contradictory*.

The first statement is tantamount to a *denial of human rationality*, thus demoting us to the level of mere animals. It is clearly unacceptable, not only because of its logical consequences in the field of religion, but even in scientific endeavors if taken in its full meaning.

The second premise denies that Faith is a rational act: one believes "just because". Whoever accepts something as true *without reasons*, without proofs, is acting irrationally. And feelings are instinctive reactions, mostly involving our body, for which we can never be responsible.

The third assertion requires the acceptance of the unique character of the Bible, as God's Word, either because *the book itself says so* or because there is a tradition holding it to be so *within a given cultural milieu*. But all that is also applicable to the Koran, and there is no objective rule to distinguish canonical books from apocryphal texts.

Finally, the rejection of the *Principle of Non-contradiction* destroys every logical foundation in any area of human knowledge.

Only the Catholic Church, relying, not on printed pages, but on the living teaching of the Apostles and their successors, *upholds human reason* in its full value. Only the Church, assisted by the Spirit, could select the true presentations of Christ's message, rejecting fables and human additions. And only the Spirit guarantees that Christ's doctrine is correctly understood and transmitted through the centuries.

APPENDIX I:
The historical evidence for Jesus (Yeshua) of Nazareth and his death by crucifixion

Non-Christian Sources for Jesus
• Tacitus (AD 55-120), a renowned historical of ancient Rome, wrote in the latter half of the first century that `Chrestus ... was put to death by Pontius Pilate, procurator of Judea in the reign of Tiberius: but the pernicious superstition, repressed for a time, broke out again, not only through Judea, where the mischief originated, but through the city of Rome also.' (Annals 15: 44).
• Suetonius writing around AD 120 tells of disturbances of the Jews at the `instigation of Chrestus', during the time of the emperor Claudius. This could refer to Jesus, and appears to relate to the events of Acts 18:2, which took place in AD 49.
• Thallus, a secular historian writing perhaps around AD 52 refers to the death of Jesus in a discussion of the darkness over the land after his death. The original is lost, but Thallus' arguments - explaining what happened as a solar eclipse - are referred to by Julius Africanus in the early 3rd century.
• Mara Bar-Serapion, a Syrian writing after the destruction of the Temple in AD 70, mentions the earlier execution of Jesus, whom he calls a `King'.
• The Babylonian Talmud refers to the crucifixion (calling it a *hanging*) of Jesus the Nazarene on the eve of the Passover. In the Talmud Jesus is also called the illegitimate son of Mary.
• The Jewish historian Josephus describes Jesus' crucifixion under Pilate in his Antiquities, written about AD 93-94. Josephus also refers to James the brother of Jesus and his execution during the time of Ananus (or Annas) the high priest.

Paul's Epistles
• Paul's epistles were written in the interval 20-30 years

after Jesus' death. They are valuable historical documents, not least because they contain credal confessions which undoubtedly date to the first few decades of the Christian community. Paul became a believer in Jesus within a few years of Jesus' crucifixion. He writes in his first letter to the Corinthians: "For I delivered to you first of all that which I also received: that Christ died for our sins according to the Scriptures, and that He was buried, and that He rose again on the third day according to the Scriptures, and that he was seen by Cephas (Peter), then by the twelve." This makes clear that belief in the death of Jesus was there from the beginning of Christianity.

The four gospels
• The four gospels were written down in the period 20-60 years after Jesus' death, within living memory of the events they describe. The events which the gospels describe for the most part took place in the full light of public scrutiny. Jesus' teaching was followed by large crowds. There were very many witnesses to the events of his life. His death was a public execution.

Manuscript evidence for the Bible and its transmission
The manuscript evidence for the Greek scriptures is overwhelming, *far greater than for all other ancient texts.* Over 20,000 manuscripts attest to them. Whilst there are copying errors, as might be expected from the hand of copyists, these are almost all comparatively minor and the basic integrity of the copying process is richly supported.

Furthermore, when Western Christians studied the Hebrew scriptures during the Renaissance, they found them to agree remarkably closely with their Greek and Latin translations which had been copied again and again over a thousand years. There were copying errors, and some other minor changes, but no significant fabrications of the stupendous scale which would be required to concoct the story of Jesus' death.

Likewise when the Dead Sea Scrolls were discovered they

included Hebrew Biblical scrolls dating from before the time of Jesus. These too agreed very closely with the oldest Hebrew Masoretic manuscripts of more than a thousand years later. Again, we find no fabrications, but evidence of a remarkably faithful copying.

Conclusion: Jesus of Nazareth is a figure of history

Clearly there are events recorded in connection with Jesus' life that many non-Christians will not accept, such as the miracles, the virgin birth, and the resurrection. However what is beyond dispute is that Yeshua (`Jesus') of Nazareth was a figure of history, who lived, attracted a following in his life time amongst his fellow Jews and was executed by crucifixion by the Roman authorities, after which his followers spread rapidly. Both secular and Christian sources of the period agree on this.

The primary sources for the history of Jesus' public life are the gospels. These were written down relatively soon after his death - within living memory - and we have every indication that these sources were accepted as reliable in the early Christian community, during a period when first and second hand witnesses to Jesus' life were still available.

The statements about Isa (Jesus) in the Qur'an were made six centuries after Jesus' death, without a shred of proof or evidence. They must be judged against the historical evidence from these first century sources, and not vice versa. And we must remember that the tenets of Islam reflect Christian beliefs (the only non-polytheistic religion in the Near East in the 6th and 7th centuries) but avoiding dogmas that were difficult to understand: the Trinity, the Incarnation and divinity of Christ, his death and resurrection, even the nature of eternal happiness. Thus it is logical that a purely human viewpoint of Christ as a great Prophet would admit his historical character while leading to the denial of anything that would imply divine nature and authority, that would be superior even to the unique role claimed by Mohammed.

Because of the frequent appeal to the Qur'an (Koran) against the facts presented above, it may be useful to check more in detail the following web pages:

http://www.debate,org.uk/topics/theo/islam_christ.html
http//ww.debate.org.uk/topics/theo/qur-jes.htm
http://www.answering-islam.org/Intro/replacing.html

Further reading: *The Jesus I never knew*, by Philip Yancey.

APPENDIX II:
The historical value of gospel narratives

We have insisted upon the historical value of testimonies given, orally and in writing, by those who lived with Christ and heard his teachings: *only thus can our faith be rationally acceptable.* But it is this historical character that is denied even by Christians outside the Church, particularly by the Protestant critics and exegetes of the last 150 years (Strauss in 1835, Renan in 1863, Bultman in our time). Their denial, especially concerning Christ's miracles, stems from two sources:

- First, it is said that the Gospels are a late compilation (put together in the second part of the 2nd century) of oral traditions beginning with the preaching of the disciples, later embellished and reworked by Christian communities during the first hundred years after the death of Christ. We do not have testimonies of those who really lived with him.

- Second, the Gospels are classified as belonging to the type of symbolic literature where mythical meanings override real data, following patterns common to poetic and marvelous stories typical of the ancient Near East. The only real information that can be obtained from the Gospels is the fact that the Christian community of that time shared the conviction that Christ was God's envoy to fulfill his promises to Israel and bring salvation to the world.

Both arguments are alike in presenting preconceived positions with very little scientific support. Against the claim that the Gospels were written rather late, we can now counter that the fragment 7Q5 of a papyrus from Qumram, dated with certainty around the year 50 A.D., has been shown (in 1972) by Fr. José O'Callaghan S.J. to contain a few words of the Gospel of St. Mark, c. 6 verses 52-53. Nobody has objected with any *scientific* argument against the identification or the date, even if many refuse to accept what is incompatible with their tenets.

In 1996 Carsten P. Thiede, a German Protestant scholar,

identified fragments of the Gospel of Matthew (7 verses of chapter 26) in a papyrus kept at Magdalen College (Oxford) and found at Luxor, in Egypt. Other fragments, probably from the same source, are kept in Barcelona (*Fundación San Juan Evangelista*). Both texts can be dated, by the type of writing, calligraphy and abbreviations, to the period around the year 60 A.D. Details of this work can be found in Thiede's book *Eyewitness to Christ* (Doubleday, N.Y.) Once more, negative reactions were published very soon. But we can say that there is an archaeological proof of the fact that both Gospels were known in Christian communities of outlaying areas of the Roman Empire when eyewitnesses of the activities described in the Gospels were still alive.

An obvious indication of the early dates for the Gospels can be mentioned: If they were written after the destruction of Jerusalem (70 A.D.) the writers would have had the best opportunity to comment that such destruction was the fulfillment of Christ's prophecy, thus strengthening their case about his divinity. Not a word is found in any of the New Testament books to suggest that the writers were aware of such an important event!

Regarding the type of literature, recent studies show a perfect correspondence between Greco-Roman biographies of that period and the Gospels. In 1978 Talbert classified them within the biographical genre; Schuler in 1982 asserted the biographical nature of Matthew's Gospel. In 1984 Cancik (ed.) did the same with the Gospel of Mark. In the book *"Ascent and Decline of the Roman World"* Klaus Berger (also in 1984) shows how the Gospels are very close to the "lives" of ancient philosophers. In 1992, Burridge stated that the present trend to consider the Gospels as true and proper "lives of Jesus" is reasonable, when comparing them to 10 Greco-Roman "lives" written between the 5th century B.C. and the 3rd A.D. (See the details in the book –in Spanish- *Jesucristo, Salvador del Mundo*, BAC 1996)

Other books worth reading are, by Vittorio Messori, *Hypothesis about Jesus, He suffered under Pontius Pilate* and *They say that He rose from the Dead.* They critically discuss the historical nature of the Gospels. The same author wrote also *The Miracle of Calanda.*

3. Science & faith: chance and design
by Manuel M. Carreira, SJ.

There is a far ranging debate, in biological and theological environments, about the opposition between the Darwinian view of evolution –from first cell to Man- and the tenet, most frequently attributed to American evangelical fundamentalists, that insists upon an intelligent design for a freely chosen purpose. I would like to offer the viewpoint that neither chance nor finality are scientific presuppositions when talking about evolution, and also that there is no need to choose between them as if they were *contradictory* (saying "yes" and "no" about the same subject from the same viewpoint.)

First of all, I would like to remind my readers of the limits of scientific methodology. "Science", in the sense in which we now use the word as describing a type of knowledge that is different from the "Humanities", deals with *the interactions of matter.* They are ways of acting that can be checked by some experiment, leading to measurements that can then be used to predict behaviour in the future and to infer previous states of a given system. This is what we mean when we attribute to science the objectivity and repeatability that allows for the same result to be obtained by any researcher of any time or culture. "Irreproducible results" are not accepted as evidence, and no theory can be given scientific status if it cannot –in principle, not due to technological limitations- lead to a proper experimental check. The theory might be mathematically and conceptually very appealing, but it will remain as "science fiction" if it can never be tested.

As an example of the logical consequences of this methodological viewpoint, it should not be considered a true *scientific* statement the assertion that an infinite value is attained for any physical parameter in any given situation. A process can have no *logical* limit –for instance the collapse of matter within a black hole- but it can never be said that after a given amount of time the density will be infinite: no possible instrument can truly measure an infinite value. For the same reason, if we were to accept an infinite density at the beginning of the Universe, it would be impossible to calculate any finite value after a microsecond or any other time interval.

"Other Universes", frequently postulated as a way out of difficulties arising in our physical description of the Universe we are in, *by their very concept* are totally unknowable, no matter how many equations suggest that they *might* exist. They are science fiction, and a very poor way of sweeping under the rug the problems that we cannot solve in the only Universe that we know and test. And it is a totally gratuitous assumption to state that whatever is mathematically possible *must* exist: mathematics is a human language useful to describe reality in its quantitative relationships, but it is only a language, not an imposition upon nature or a magical incantation to make things happen.

"Finality", "chance", "purpose", *cannot be detected by any experiment or reduced to a number in an equation.* Even the most obviously purposeful product of our technology cannot be *scientifically* proven to exist *for a reason.* But we constantly *infer* purpose —finality- from the study of the properties that the object has, and from the logical deduction of its unsuitability for it if those properties were significantly altered. An automobile would make no sense if it had triangles instead of circular wheels, or if it lacked a steering mechanism, or if it were made of a very brittle material, and so on. It is obviously not meant to move through the ocean or to fly, or to grow plants within its volume; it only makes sense if it is made to move over a fairly even and hard surface, under human control.

Human *thought* (not just the *activity* of neurons when we think) cannot be experimentally checked; the same applies to the possible value of an idea when writing a poem. In fact, neither the literary value of any book nor the artistic level of a painting can be experimentally proven. This is also true of our family relationships, our sense of duty, of our social concerns: *all that constitutes truly human life and culture is impossible to detect and quantify following strict scientific methodology.*

In science itself, the ultimate questions cannot be answered by any equation. In the words of John Archibald Wheeler[1], the most important one is "why is there something instead of nothing". *This lies outside of Physics*, to be the object only of a Metaphysical inquiry. Or, as Stephen Hawking admits, the equations describe a Universe, but they cannot explain *why* a Universe exists that follows the equations[2]. Time and again the ultimate "why" and "what for" of matter itself appears as an unsolvable mystery if we restrict our viewpoint to the experimentally provable facts.

But we are not satisfied with the limited data and methods of science. During the second half of the past century, renowned cosmologists have proposed the "Anthropic Principle"[3], which tries to infer the purpose of the Universe from a detailed study of the consequences that would follow, according to physical laws, from changing any parameter of matter, already at the very beginning of its evolution. It must be said quite clearly that the quest for an answer requires leaving the purely physical way of thinking: it is a *metaphysical* principle that is being developed from the data of the sciences of matter. But it is quite logical to be interested in an answer, just as we are interested in the beauty of a poem and not only in the chemical composition of the paper and ink of the book where we find it.

Wheeler presents the Anthropic Principle as indicating that the Universe points to intelligent life as the only satisfactory reason for the choice of parameters at the Big Bang itself. For that, he goes to the very core of the nature of matter as studied by science: it is "adjustable" to exist in many possible ways, since it is in constant change. And whatever can exist in different ways needs to be *extrinsically* determined to exist in a particular form rather than any other. We could say that the most universal statement about material beings refers to their dependence upon *time*, since any change implies different properties in successive moments. We are thus led to the need to accept either causality towards the past –Wheeler's proposed solution, which is the ultimate example of circular reasoning- or the creation of matter by a non-material agent, *not tied to time or space*. But true creation *necessarily implies an infinite power* [4] -with knowledge of all the unlimited possibilities of making a Universe- and a choice of parameters for the one that is actually created. *Such a choice implies a purpose.*

It would be absurd to consider as a sufficient reason for creation -by a non-material, personal Being- the simple existence of stars burning themselves out for billions of years, or the crawling of mindless biological entities on the surface of some planet. Physics finds that the most stringent limits on physical parameters are imposed by the existence of intelligent life at least in one place of the Universe; both Philosophy and Theology concur by stating that the only logical purpose of a personal Creator must be the existence of other personal beings –intelligent and free- that can know their debt of gratitude and partake of the happiness of the infinite source of being who wants to share its very life.

From the viewpoint of Physics we can also establish the limits of material activity. Modern science accepts only four interactions –gravitational, electromagnetic, strong and weak nuclear- that *define* what matter is with an operational definition, typical of scientific methodology. None of those interactions includes *consciousness, abstract thought or free will* among its effects, thus leaving outside the realm of matter the most obvious activity of Man. If the Creator –personal, intelligent and free- is non-material, it is logical to accept the *possibility* of creating other non-material entities endowed with similar capabilities, even if at an infinitely lower level: only what is above matter can act in a way that surpasses the four interactions that define matter itself.

Cosmic Evolution

When the Universe springs into being, without any previous state from which to derive its properties (before the Big Bang *there was no before*) any initial set of parameters can be considered "arbitrary" in the obvious sense that there is no previous reason for it to exist instead of any other possible one. But a purposeful Creator -not limited by time constraints (proper only of matter)- *can and must* choose the initial conditions with full knowledge of the consequences for all time of making the Universe in a particular way, down to the most intimate nature of every particle and quantum of energy, and their activity *at each moment of cosmic evolution.* There can never be something unexpected or surprising for the Infinite Mind that sees all of cosmic history as present in its eternal "now". This *does not* imply that at each turn the Creator is *imposing* with a "fiat" what each atom does: having created matter with properties that are suitable for the intended purpose, matter acts according to its laws, due to its nature.

This viewpoint is misinterpreted when it is construed as

42

denying human freedom. If I were able to build a time machine -not to travel to the future, but to observe it in a TV screen- I would *know* what will happen (both as a result of physical laws and of free human choices) but my knowledge would not *determine* that what I observe will happen. The same could be applied to God's total knowledge of the future as present: for God there are no waiting periods or unforeseen developments. And nothing –even knowledge- can be added as a result of cosmic evolution to the truly infinite Being, *without possibility of change*, as it is logically necessary to describe its nature when its existence is not limited by space or time. This is something that some recent theologians seem to forget when they propound a "Process Theology" in which the Divinity evolves and becomes more perfect by some kind of feedback from its own creation.

From the first instant of the Big Bang to the present, nature develops structures that lead to the synthesis in massive stars of the elements necessary for life with its amazing complexity. When this evolution reaches the point of forming here on Earth a suitable environment, life appears, in spite of the infinitesimal probability of its happening[5]. Billions of years later, in an unpredictable series of small changes and catastrophic extinctions, living matter is finally ready for its role as partner of the human spirit in the unique "rational animal" that is Man. But matter can only evolve into new forms of matter: is thought and consciousness due to matter alone?

Biological Evolution

When we find that primitive life, as recorded in sedimentary rocks from billions or millions of years ago, was limited first to microscopic single cell organisms, and then developed into more and more varied and complex forms, it is illogical to deny evolution *as a fact*. Most species have become extinct through the ages, with the well known death of the dinosaurs 65 million years ago freeing our planet from their oppressive presence and thus opening the way for the full development of mammals. Only an obsessive reading of the Bible as a literal treatise of geology and biology (which had to be compatible with an apparent abundance of contradictory evidence) would imply that evolution never occurred.

But the philosophical and theological problems in this field are not due to the facts that science presents with undeniable strength. The crux of the controversy lies in the added *philosophical* presuppositions regarding two key points: the driving force and mechanism for evolution in general, and the step from non-intelligent primates to Man.

In the first case, the battle lines are drawn in terms of an "either-or" choice: evolution has happened either *by chance alone* (followed by the survival of the fittest and adaptation to the environment) or by an *intrinsic drive impressed by the Creator* according to an "Intelligent Design" where nothing happens by chance. It is here where a more nuanced use of words and better defined concepts become necessary to avoid extremes that might end up denying either evolution itself or even the existence of the Creator.

"Chance" is not a *measurable* parameter of matter, like electric charge or mass. It cannot be put into an equation as a factor (even if the related concept "probability" can be used to compute an expected outcome.) In a more pedestrian way, one could say that "chance" is only a more polite term to reply to a question for which we do not have an answer, instead of saying "just because". Why does a particular squirrel cross in front of my car today and gets killed? There is no reasonable way to establish a *predictive correlation* between my driving and the constant running around of the squirrel, and thus the answer is that the death of the animal under my car is a matter of chance. The same can be said of a cosmic ray with a given energy hitting a particular chromosome in a reproductive cell of a particular animal and causing a mutation. This is *chance*, and only chance, as seen from the viewpoint of science. In this sense, chance is a common element present in our life and in most of the independent happenings in the Universe. But each material interaction is a *necessary* consequence of the properties and forces present at each moment: there is no room in Science for any kind of "spontaneity" or "creativity" that would logically imply a degree of free will even in the most basic particles of matter. To say otherwise, besides being totally gratuitous, would undermine the possibility of certain predictions, and science would be impossible.

But from the viewpoint of the infinite Mind that sees every detail of the entire cosmic evolution, from the Big Bang to its last gasp, there is never an unexpected happening. Chance cannot apply to the perfect foreknowledge that the Creator has, and *that makes the choice of initial conditions and laws of development a sure way to obtain the finality for which creation is meant.* There is no opposition or irreducible "either-or", because *chance* and *design* are asserted at two different levels, and *neither of those assertions is a subject for experimental verification.* Scientific methodology *cannot prove finality or its absence*: in both positions we are introducing philosophical considerations to give *a reason* for facts that are accepted as such, but the reason cannot be found in the laws of nature –thus the appeal to "chance"- while Philosophy and Theology do provide an acceptable answer.

The controversy regarding the teaching of evolution in school is misguided. Biology should cover the multiple stages of life on Earth, and explain the *mechanisms* through which mutations occur and are eventually such as to give rise to the present variety of species and individuals in all living forms. This is still a largely unfinished task, where eminent biologists are still baffled by their inability to explain very drastic evolutionary steps in terms of small changes[6] and to attribute to the same environments (for instance, in the sea) the enormous range from single cells to animals like the octopus and the whale. The same can be said about the coincidence in the same organism of all the genetic changes needed for a significant evolutionary step, and the need for many individuals so changed to make it last.

45

But the question of chance or design does not touch anything that biology has to deal with. That is a matter to be properly discussed in a Philosophy course, perhaps within the larger context of Cosmology and the reason "why there is something instead of nothing". And in this approach, and only here, it is proper to discuss if the Universe has a purpose, both as a single complex entity with multiple levels of structure and activity, and in the individual happenings which Science necessarily attributes only to the necessity of physical laws and to the chance coincidence of unrelated events.

The Origin of Intelligence

Once more, the popular way to present the problem of Man's origin is to state an "either-or" alternative: either by *simple evolution* from a primate, common ancestor of apes and Man, or by *direct creation* from inanimate clay to form the body (unrelated to other animals) where a spirit is then introduced. This is also construed to mean that in the first alternative there is no soul, just highly structured matter, while in the second one the soul is the defining element for the specific nature of Man as a "Rational Animal" and also the "Image and Likeness" of his Creator.

The basis for a reasonable solution to the problem is found in the concepts of *matter* and of *intelligence*, both frequently used *without the required first step of a suitable definition of each*. No serious discussion is possible unless we first clearly state what we mean by each term, and the starting point cannot be an *a priori* philosophical position accepting *only* the existence of matter and explanations of reality based upon its activities.

As stated earlier in this essay, matter is defined in Physics in terms of its *interactions*. We do not have an intuitive grasp of essences, even for the most common objects in our daily experience, but we attribute patterns of activity to the nature of things, and thus we identify different objects by what they do. In the popular way of saying it, "if it looks like a duck, and waddles as a duck, and quacks as a duck, it is a duck". O, in more scientific terms, if it has the mass of an electron, and the charge of an electron, and the spin of an electron, it is an electron.

We accept four ways of acting that can be checked experimentally, and thus we state their effects and range and specify the objects subjected to each of them. The totality of the material world –particles, energy, the physical vacuum, space and time- are affected by gravitation, which causes attractions and distortions we need not describe in detail. Particles with something else, "electrical charge", experience very strong attractions and repulsions that explain common properties like rigidity, hardness, chemical reactions, living structures. The two nuclear forces explain the atoms of the elements in the periodic table, their formation inside stars and the relative abundance of each element through cosmic history. There has been no need, as yet, to invoke a fifth force to cover any of the multiple happenings in any level of the material world. And the most basic law of Science states that in any interaction, *"matter is never created or annihilated; just transformed"*, even if the change can be as drastic as the disappearance of particles in the form of pure energy or the other way around. Other "conservation laws" (of net charge, of linear and angular momentum…) also restrict the way matter develops in our experiments.

If we look for an explanation of Man, we cannot stop at this level. The most evident fact to each of us is *our own consciousness*, the awareness of thinking processes and of free choices. Chemistry, Physics and Biology can give a detailed description of all the energetic changes that take place when I bend my arm, but they cannot say *why the arm bends when I want, and why I am conscious of bending it by a free choice*. This consciousness, if due to the material activity of neurons in the brain, should logically contain –first of all- a clear awareness of such activity. But nobody knows even the existence of neurons unless the fact is learned in Biology and Anatomy: matter is not conscious of itself for anything internal to our own body. In the act of seeing, I am quite aware of the object I see, but not of the changes in the retina or the processing of signals in the brain. *Consciousness cannot be attributed to matter according to its operational definition: none of the forces accepted by physics gives even a hint of being adequate to produce it.*

What is *intelligence?* We loosely talk about the intelligence of a dog or a dolphin, because they can be taught tricks, by training that produces a specific response –a way of acting- as a reaction to some stimulus (a command or a gesture.) More inexact still is the term when applied to computers, that have no interest in knowing anything or satisfaction in performing any computation, purely determined by an external programmer. But intelligence is not a way of acting, either by instinct, by conditioned reflexes or by electronic commands: it is a way of *knowing*, even abstract concepts that cannot enter the mind through the senses. Philosophy, pure Mathematics, the most modern scientific theories, are so far removed from our sense experience that even imagining their contents becomes impossible. We deal with aspects of reality that cannot be tied to our daily experience, and we value their logical beauty, their cogent necessity in abstruse proofs, their pure reasonableness. From Euclidean theorems in Geometry to the intricacies of Superstring theory, true intelligence is far removed from the realm of the four interactions of our experiments. This makes totally arbitrary and illogical the effort to attribute cultural developments of any kind to the blind actions of particles and energies.

No scientist will accept claims of controlling an experiment by thought or will power: a proof of the deep-seated conviction that our thoughts and desires *do not add anything* to the material environment described by Physics. The same attitude is reflected even in the wrong efforts to reduce human personality to a genetic code –totally invariant from birth to death- as if all the achievements of the greatest genius in science, art or literature were nothing worth mentioning. This is the contradiction of trying to reduce everything to the activity of matter, when it is obvious that *science itself cannot be identified with any atomic arrangement or series of energy changes.*

We are, consequently, faced with the undeniable double level of human activity, which requires two different sources, matter and spirit, intimately joined in a personal unity that is the subject of both, with mysterious mutual influences, but with clearly different functions. It would be absurd to deny our materiality, but one could say that it would be more illogical to dismiss our non-material consciousness. It is the search for Truth, Beauty and Goodness that drives the best efforts of our human existence, from the Stone Age caves to the present Space Era.

Trying to reduce intelligence to matter, it is claimed – without any proof- that when matter in the brain is sufficiently structured, intelligence *arises* or emerges spontaneously as a new level of activity, but without any new element being present. This is totally void of explanatory power, even if it is reinforced with the example of modern electronics, where electrical currents are the only provable content of a TV image or the display of a computer screen. The show in my TV will not be entertaining or boring as a result of the quality of the electrical currents: we rather appreciate or blame the work *of some mind* in preparing the script and choosing the program contents. To say that there is nothing more in TV than a flow of electrons would be as superficial as to equate a Shakespeare play with the black stains on the pages of the printed version, to determine its literary value..

If matter, even with the highest degree of order and structure in the brain, cannot produce thought, then it is impossible to attribute intelligence to the development through the ages of primates with higher and higher volumes of cerebral tissue. This is also incompatible with the fact that dolphins and elephants have more brain than we do, and hydrocephalic humans with extremely reduced cortex show no impairment of their minds. Man cannot evolve by simple material changes in the genetic programming of a previous non-human: thus the theory of evolution is falsely presented as a *sufficient* answer for our existence.

It is true that biology must provide the suitable basis for the human spirit —and there is nothing shameful in using living tissues instead of dead clay for the purpose- but the human spirit can only be created by the spiritual Creator who is also the reason why the Universe exists. To say otherwise is not only poor Philosophy, but also poor Science.

Science and Christian Faith

We have reached a point where we have to accept that neither *Science* (Cosmology, Chemistry, Biology) nor *Theology* provide a complete answer to the "why" and "how" of the existence and evolution of the Universe and of ourselves in it. Both are limited ways of knowing a marvellous reality that surpasses our understanding at almost every step, even if data are always to be accepted, no matter how difficult it might be to include all of them into a coherent picture. We need the complementary viewpoints of different methodological approaches, and we have to be on guard not to introduce philosophical presuppositions or prejudices of any kind as a basis for what is presented as a scientific or theological assertion.

Whenever a Science-Religion conflict has arisen during the past history of human thought, its roots can easily be found in the unconscious attempt to reduce all forms of knowledge to a single methodology, be it the experimental approach or the literal reading of the Bible. We should all learn from past mistakes and clearly state the meaning of key terms and the viewpoint from which each problem is presented, with its limitations and strict proofs.

It is also quite evident that being a great theologian does not confer a special authority to speak about Science, any more than knowing a lot about Physics will entitle a PhD in that field to pontificate about questions of God's existence or the purpose of the Universe and of Man. If this was always true, even in past ages when it seemed possible for a person to encompass all of human knowledge, it is patently clear today, when each specialist feels at home only in a small portion of any given science or theological study.

All Wisdom is found eminently in the Infinite Being who speaks to us both with the book of Nature and with the words of Revelation. Being the absolute Truth, Order and Beauty, the Creator is also the ultimate foundation of the possibility of Science itself: in the words of Einstein, Science is possible because the world is not absurd: contradictions cannot exist between any two truths of any kind: both are partial reflections of the one Truth found in God.

REFERENCES

Wheeler, J.A. *The Universe as Home for Man.* The American Scientist, Jan-Feb 1977

Hawking, Stephen, *A Brief History of Time,* Bantam Books, New York 1988, p. 174

Barrow, J. and Tipler, F., *The Anthropic Cosmological Principle,* Clarendon Press, Oxford 1986

Using the symbols of mathematics, the only way to obtain a number (any number) from zero would be to multiply it by infinity. Neither zero nor infinity can be used to refer to real things: strictly speaking they are not numbers. No finite power can produce something from nothing.

See in *The Anthropic Cosmological Principle*, p. 565: the odds for assembling a single gene by chance are of the order of 1 in 10^{109} to 1 in 10^{217}. The number of atomic particles in the Universe is considered of the order of 10^{90}, about a trillion-trillion times smaller. For the entire human genome, the odds are impossible to even imagine: 1 in $10^{12\text{ million}}$.

Behe, M.J., Dembski, W.A., Meyer, S.C., *Science and Evidence for Design in the Universe,* Ignatius Press, San Fco. 2002, p. 113-128

4. Scientific critiques of
materialistic Darwinism
by Julio A. Gonzalo

In three millennia of cultural history the most prominent philosophers (Plato, Aristotle, Agustin, Thomas Aquinas, Suarez) have been rather on the side of "theism" than on the side of "skepticism" (or "atheism"). All of them are "realists". Unlike Kant and his followers, who are "subjectivists", and then more akin to skepticism (either pantheistic or atheistic).

Curiously, the two greatest physicists[1] of the 20th century, **Max Planck** and **Albert Einstein** were "objectivists", no doubt unfriendly, in their mature years, to the rather crude world views of Ernst Mach. Few years before his death, Einstein, in a letter to his good friend Maurice Solovine, says:

"Even if the axioms of the theory are posited by man, the success of such procedure (Newton's gravitational theory) supposes in the objective world a degree of order which we are in no way entitled to expect a priori. Therein lies the "miracle" which becomes more and more evident as our knowledge develops… And here is the weak point *of positivists and* professional atheists, *who feel happy because they think that they have preempted not only the world of the* divine *but also of the* miraculous… (emphasis added)."

(For more details, see discussion of S. L. Jaki, Ref.1 below)

Kant's claim that the *universe* was a bastard product of the metaphysical cravings of the human intellect, put forward in good measure to discredit the classical *cosmological argument* to prove God's existence, was denied with deeds by both Planck and Einstein.

Science, in particular Darwinian Evolutionism, a rather "soft" Science in comparison with Physics, Geology or Astrophysics, is used today as a powerful weapon to discredit *Religion.* But Religion is nothing else but a quite natural disposition in man to recognize a *"created order"* in the world around. The observable *"Intelligible Design"* in nature and the fact of *"intelligent observers"* perceiving it, obviously, presuppose an *"Intelligent Designer".*

Modern *Science,* far from making "unnecessary" a Creator, is *a monumental witness* to the *intelligibility* of a *finite, open* and *contingent* universe. Exactly the opposite to what the propagandists of materialism are claiming today.

Defenders[2,3,4] of ID are, in my opinion, basically right; when they say that modern science is *proof* of *"intelligible design"* in nature. This implies, of course, an Intelligent *Creator.*

Beginning with Michael Denton and Philip E. Johnson (two distinguished precursors of the ID movement) a representative sample of scientists, including conservative Catholics, conservative Protestants, and orthodox Jews (all of them competent defenders of ID) is given below. Their arguments may not be always apodictic, but their inferences are basically sound in my opinion.

Michael Denton (Australian Molecular Biologist) author of *"Evolution: A Theory in Crisis"* and *"Nature's Destiny"*. In his first, pioneering book, he shows that "evolution", the key stone of our modern world-view, as initially proposed by Darwin, is coming increasingly under fire from many sides.

Although, according to Denton, the thesis of orthodox Darwinism relative to the emergence of *new species* seems to be correct, its claims about the global interrelationships of *classes* and *orders*, and especially about the *"origin of life"*, are based upon very shaky assumptions.

In "Nature's Destiny" (his latest book) Denton shows that the scientific evidence is consistent with the traditional view that man is the fundamental end and purpose of the whole physical cosmos, and that the actual historical sequence of scientific and technical discoveries points to a pre-arranged order in nature.

As noted by **Michael Behe** (about whom more below) in the back cover of "Nature's Destiny", Denton shows that the universe was *intentionally designed* for *human beings*: "From the laws of physics to chemistry to biology, from the properties of water to the characteristics of fire, he shows the goal of the cosmos to be human life. The scientific and theological consequences of this study are immense".

Philip E. Johnson (Graduate from Harvard, Emeritus Professor of Law, university of California, Berkeley). Author of *"Darwing on Trial"* and *"Defeating Darwinism by Opening Minds"*: He combines a broad knowledge of basic biology with the penetrating logic of a leading legal scholar. His first book is a brilliant attack on Darwin's materialistic evolution. "What if "evolution" is a word that covers up scientific ignorance of how the wonders of the living world could have been created", asks Johnson. He shows that , as the difficulties of Darwin's theory pile up, Darwinists have clung to their theory out of fear of religious fundamentalists. In the process, they have turned belief in Darwinism into their own religion.

Michael Behe (Department of Biology, Lehigh

University). Author of *"Darwin's Black Box"* and *"The Edge of Evolution"*. He does a brilliant job at explaining and illuminating one of the most disturbing problems in biology: the origin of the *complexity* that permeates all life around us. He chooses as the answer to this question one that falls outside the precincts of science proper: an *intelligent designer*. The book may be taken as a clever and original presentation of the design argument. For instance, as Behe points out, the *bacterial flagellum*, needed by bacteria to swim (of considerable medical interest therefore) has three independent parts – paddle, hook and motor- and is therefore *irreducibly complex*. Gradual evolution of the flagellum faces therefore *tremendous obstacles*. The professional literature on the flagellum is very large: thousands of papers each year. But the *evolutionary* literature on the subject is completely missing, Behe notes. According to **Peter van Inwagen**, on the back cover of the book, if Darwinians respond to this important book by ignoring it, misrepresenting it, or ridiculing it, that will evidence that Darwinism today functions more as an ideology than as a scientific theory.

William Dembski (Research Professor of Philosophy at South Western Baptist Theological Seminary, Forth Worth) has doctorates in mathematics, engineering and philosophy and is author of *"The Desing Inference: Eliminating Chance Through Small Probabilities"*, and co-author of *"The Design of Life: Discovering Signs of Intelligence in Biological Systems"*. Dembski is the proponent of the so called *"explanatory filter"* to prove *"intelligent design"*: (1) One asks *first* whether the observed phenomenon has a *high probability* of occurrence. If so, it can be attributed to a natural cause, like gravity (which, however, might be the result itself of "intelligent design"). If not, it could be due to something else. (2) If the phenomenon in question, on the contrary, is *highly improbable*, it might be the result of *many independent trials* (something of the order of one in 10^6 after say 10^4 to 10^8 trials) and therefore it can be "explained" by chance. (3) Finally, if the probability of occurrence is *"increasingly small"*, something like one in 10^{50}, with only a comparatively reduced number of trials, one is entitled to *infer "intelligent design"*. For instance, there are about 10^{12} galaxies in the universe, each with about 10^{12} stars. If one were able to show that the probability of finding a planet capable of sustaining human life is only of the order o one in 10^{50}, then one would be entitle to conclude "intelligent design".

(In fact, **Paul Dirac**, one of the co-founders of Quantum Mechanics (by his own admission, an agnostic in religious matters) pointed out that if one could show that our planet is the only one housing intelligent life, that alone would be a conclusive argument for the existence of a Creator).

Stephen Mayer (Senior Fellow if the Discovery Institute, Ph.D. in history and philosophy of science from Cambridge University) is the author of *"Signature in the Cell: DNA and the evidence for Intelligent Design"*. He did his thesis dissertation on the history of origin of life biology, and on the methodology of the historical sciences. He has undergraduate degrees in Physics and Geology and is contributor to "God and Evolution", edited by Jay Richards (Discovery Institute Press). As pointed out by Mayer, "Theistic Evolution" as put forward by some well-meaning Christian Darwinists has problem: (1) There is no compelling reason to assume that the "Designing Intelligence" responsible for life in the universe has *confined* his activity only to the very *beginning*. (2) There is no compelling reason to say that purely *material* processes and mechanisms are *sufficient* to account for the *origin of life,* and for all the major steps along the history of life in our planet. (3) There is no reason to assume that the enormous amount of *biological information* discernible in the living world came out suddenly from nowhere. It has been said that "DNA is like a computer program, far more advanced than any software ever created". If so, *how* did the information in the DNA arise? DNA shares the assembly instructions for building the many crucial proteins required to make up living cells. But the DNA itself is made up in cells only. As Mayer points out, the laws of nature can *transmit,* but cannot *generate* information. Was the *information* necessary to produce a *DNA molecule* present just a very small fraction of a second after the *beginning* of the universe when the size of the universe itself was much smaller that the smallest elementary particle? The only answer is that the presence by pure chance then and there of such information is utterly inconceivable.

Jay W. Richards (Senior Fellow of the Discovery Institute, Ph. D. in philosophy ad theology from Princeton Theological Seminar and B.A. with majors in Political Science and Religion). Author of *"Money, Greed and God"* and co-author with Guillermo González of *"The Privileged Planet"*, Editor of *"God and Evolution"*[7] (Discovery Institute Press, 2010). After reviewing the views of classical defenders of the Design Argument, he points out that, perhaps, even if some of their arguments are partially dated or incomplete, they can be taken as complementary rather than opposite to recent ID arguments. He quotes **William Dembski** as saying "I am much bigger fan of the Church Fathers than of William Paley. I like Paley and think he has a lot of good insights. But I think the watch metaphor was in many ways unfortunate…". Even today, there are obvious "gaps" in the most rigorous man made scientific theories, physical, chemical even mathematical, not to say biological.

The universe is "intelligible" albeit *not fully* intelligible for us. The universe is *not* well represented by a clock, of course, but in our solar system, in our galaxy, in the totality of matter and radiation which makes up our finite universe, there are obvious "clock like" features. God's creation on the other hand includes men, endowed with intellect, will and freedom. As shown by Gonzalez and Richards, our "Privileged Planet", as well as the universe, is a book in which intelligent observers can investigate and read. Nature is meant by its Author not only for *life* but for *discovery*. Therefore *contemporary science* is a monumental *proof* of an Intelligent and Omnipotent *Creator*. When atheists reject God's existence, they are obviously inconsistent: "*If there were no God, there would be no atheists*".(G. K. Chesterton)

Of course the arguments for an *intelligently created* world can always be improved, clarified, made more simple, but, basically, they are *common sense* arguments sometimes formulated in scientific terms and in a language which is more familiar to modern men.

Richards quotes Cardinal Joseph Ratzinger (before he became Benedict XVI) suggesting a partial analogy between the Aristotelian *substantial form* and the modern scientific notion of *"information"* as embodied and encoded in the DNA.

Good science complemented by good philosophy (and revelation) do not need to be on the defensive in its present confrontation with modernism, materialism and secularism.

Jonathan Wells (Senior Fellow of the Center for Science and Culture at the Discovery Institute, A.B. in physical science from the University of California at Berkeley, Ph.D. in Theology from Yale University and Ph. D. in Biology from the University of California at Berkeley). Author of *"Icons of Evolution"*, *"The politically Incorrect Guide to Darwinism and Intelligent Design"* and *"The Myth of Junk DNA"*. In his polemical book "Icons of Evolution" Wells pinpoints *ten* problematic presentations in biology textbooks which tend to support a materialistic interpretation of evolutionary theory:

1st) *Life from non-life*: The famous *Miller-Urey experiment* does not work, according to Wells. It seems that nowadays Miller himself agrees.

2nd) *The tree of life:* Wells admits that the field is overflowing with evidence, but he says that conflicting evidence has engendered rampant confusion leading to *"tree diagrams"* that resemble rather *crisscrossing "thickets"*.

3rd) *"Homologies"* of structure (similarities of bones in the vertebrates limbs, of amino acid composition in widely separated taxa, etc): According to Wells they are often not unquestionable evidence of descent from common ancestors; sometimes they are presented in such a way that look like *arguing in a circle*.

4th) *Haeckle's drawings* of vertebrate embryos: Wells asserts that they are forged, and that even most biologists sympathetic with Darwin's have known it for a long time; and that, until *very recently*, most Darwinists have said nothing about them.

5th) *Archaeopteryx:* This fossil is commonly cited as proof of evolution, a "missing link" between dinosaurs and modern births. Wells argues that Archaeopteryx is not an ancestor of modern birds, and therefore, no proof of evolution.

6th) The celebrated case of the *peppered moth*: Wells points out, among other things, that the moths were glued on tree trunks to take the photos. So the "industrial melanism" phenomenon, for which these moths are famous, is *no support* of Darwinist natural selection.

7th) *Darwin's finches.* Cited as a case of natural selection in the wild: Wells points out that population changes in beak morphology represent a case of microevolution (cyclical variations in tweaking structures) which has never been questioned by ID proponents.

8th) *Fruit flies with four wings*: As noted by Wells, cases of four-winged flies appearing spontaneously in "Diptera" (two wings) is not, as claimed in some books, evidence for a neo-Darwinian mechanism of evolution.

9th) *Fossil horses and directionality* in evolution: Wells points out that mostly said and taught (illustrated in textbooks) on the evolutionary lineage of horses (a paradigmatic case) is wrong. Actually Wells favors of a branching description rather than the straight line of so called orthogenetic evolution.

10th)*Hominid evolution into humans*: Wells notes that the gap between anthropoid apes and humans is not conclusively filled. It was filled for many years (from 1912 to 1953) with a successful fraud ("Piltdown man") which contributed to confuse the whole issue. The cases of the "Java man" and the "Peking man", according to Wells, were not examples of scientific integrity either. Today there are disagreements between experts about which fossil species are ancestral to others. Paleoanthropology, according to Wells, is much less trustworthy than often presented.

It is a well known fact that there was a geological time when there was no life on the earth surface, and that life begun to left vestiges of its presence at a later time. This does not mean, however, that Darwin was right regarding the exclusive character of the mechanism he gave as the only exclusive explanation of evolution.

Darwin said, as pointed out by Michael Behe:

"If it could be demonstrated that any complex organ existed which could not possibly have been formed by numerous, successive, slight modifications, my theory would absolutely break down".

David Klinghoffer (Senior Fellow at the Discovery Institute) Author of "The Lord will Gather me In: My journey to Orthodox Judaism" and "The discovery of God: Abraham and the Birth of Monotheism". He received his A.B. magna cum laude in comparative literature and religious studies from Brown University. According to Klinghoffer orthodox and tradition-minded Jews are committed to a religious worldview that stands at loggerheads with Darwinian evolution. In "Guide of the Perplexed", **Maimonides**, the most respected medieval Jewish philosopher (roughly contemporary of Averroes and Saint Thomas Aquinas) dispels any compulsion to read only "literally" the Bible, including Genesis I, which describes the earth's creation. But Maimonides rejects the eternity of the world against Aristotle for two reasons: first because it "has not been demonstrated": and, second, because it makes non-sense of Judaism.

Says Darwin[6]:

"I would give absolutely nothing for the theory of natural selection if it requires miraculous additions at any one stage of descent".

According to Klinghoffer, Darwinism reached a high point in academia in 1959 with the celebration of the Centennial Celebration of "The Origin of the Species" at the

University of Chicago. Since then, he contends, at least in academic circles Darwinism has lost ground. There is a list of more than 800 signers of a public statement professing doubts about Darwinism, including researchers at UCLA, Princeton and MIT. This does not sound like a theory which has been absolutely and unambiguously "demonstrated".

Professor Joel L. Kraemmer of the University of Chicago is quoted as saying:

"A good answer (to the question of what is the most important idea taught by Maimonides) would be that it is the idea of an orderly universe governed by laws of cosmic intelligence".

Traditional Jewish sources see in the Greek philosopher **Epicurus** the paradigm of the "heretic", the "atheist" and the "free thinker".

Part of Epicurus program is to eliminate the fear of divine justice. For him human life is a purely material affair. Even the soul is made of matter. Reality, according to him, "is all composed of atoms". The world is, therefore, due to mechanical causes and there is no need to postulate theology, i.e., purpose, design, God[9]. This achieves *its modern biological expression* in Darwin's theory, according to Klinghoffer.

He ends up quoting Psalm 135:6:

"Whatever God willed, He did-in heaven and on earth".

And he ends up with the commentary of Rabbi Bachya, made almost a thousand years:

"When a thing always acts in a certain way, this indicates that its actions are not expression of its own will... One who acts out of free will, however, acts in various ways at various times".

God has created *freely*, not out of necessity, neither randomly nor deterministically.

What ID supporters are saying, basically, is that Darwinism (i.e. chance plus survival of the fittest) in not the final and complete explanation of the origin and development of life on earth.

At the same time, ID supporters are reaffirming something that is obvious: ordinary men are able to discern an astonishing order in nature. In the realm of the stars and planets, in the realms of inner solids , liquids and gases and in the realms of plants and animals.

When Darwin wrote his main theory of evolution, the atomic / molecular character of matter was not yet well established. Genetics was in its infancy. And the fundamental role of DNA in every living cell was totally unknown. Today hundred and fifty years after *The Origin of the Species*; biologists know much more than Darwin did. And, probably, in another one hundred and fifty years, a future generation will know substantially more than the present generation of scientists.

What ID supporters are trying to say some times more competently than others is that the tree of life is not an arbitrary, random development. A purely materialistic explanation is never, a satisfactory rational explanation. To infer a Creator is, of course, a metaphysical inference. But grounded in facts, ordinary or scientific facts.

Does it mean that all the arguments used by ID supporters are all the time equally conclusive and felicitous? No.

REFERENCES

[1] See f.i. S.L. Jaki "The Road of Science and the Ways to God" (Chicago: University of Chicago Press, 1978) Chapters 11 and 12.

[2] Thomas Woodward – "Darwin Strikes back" (Baker Books: Grand Rapids, Michigan, 2006)

[3] Silvano Borruso – "El evolucionismo en apuros" (Madrid : Criteria Libros, 2000).

[4] Jonatan Wells – "Icons of Evolution…" (Washington D.C.: Reguery Publishing, 2000).

[5] Jay W. Richards, Ed. "God and evolution" (Discovery Institute Press: Seattle, 2010) p.280.

[6] Francis Darwin, ed. "The Life and Letters of Charles Darwin" Vol. 2 (London: John Murray, 1887) p.227.

[7] Frederich Copleston, S.J., "A History of Philosophy: Book One" Vol. 1 (New York: Image Books, 1985) p.405.

5. The religion of Darwinism
by Julio A. Gonzalo

Stanley Miller's spark discharge experiment re-created in
1953 what then was thought to be the primeval Earth's
atmosphere in a sealed glass apparatus.

A spark device zapped the gases (methane, ammonia and hydrogen) with stimulated lighting, and a heated coil kept the water bubbling. Miller found deposited on the glass a reddish deposition which was rich in *amino acids*.

As it is well know[1], amino acids link together to form *proteins*, and proteins are the basic stuff of *living organisms*. This was the famous "primordial soup", and, at the time, the scientists speculated that living organisms, within a short time, would be *produced* in the laboratory. It did not work that way, of course.

Leaving aside the fact that the mixture of gases in Miller's experiment was not realistic (molecular H_2, f.i., is too light to be kept within the atmosphere by the Earth's gravitational field), let us *focus* on a very specific aspect of the experiment[2]: as it was to be expected, the amino acids obtained were a "racemic mixture" of *left-handed* (levo) and *right-handed* (dextro) amino acids, in the exact proportion 50% to 50%. In other words, from the point of view of chirality (handedness) the result obtained was a truly *random* result.

On the other hand: What is actually the case with the *natural*, biological amino acids we find in all living cells, within plants and animals? They are all 100% *left handed*. Contrary to the result of Miller's famous experiment, the amino acids are *not* random at all.

To put in *quantitative terms* the improbability of such a *non random* feature observable in all living systems (*lefthandedness*), we can take a look the fossil record:

After the formation of the Earth- Moon System (presumably in a cataclysmic collision) and after the early period of heavy meteor bombardment, the surface of our planet was utterly inhospitable to life of any kind. Radiation dating of the oldest rocks at that time indicates that the Earth's crust material is about 4.7 *billion* years old.

Only about *0.2 billion* years after the Earth was about 4.0 billion years old (after physical conditions on the Earth's surface began to be minimally hospitable to life) the first imprints traces of living *protozoans* were left in old rocks.

In the following *3.8 billion* years, all the way with an atmosphere already hospitable in principle to life, the likelihood of formation of similar protocells, could be assumed to be as high as during the previous 0.2 billion years.

Consequently, since the time the Earth was 4.0 billion years old, events quite similar to the event in which the first protocell was spontaneously synthesized could have taken place, not once, or twice...but roughly *twenty times*.

Therefore if the unknown process by which a primitive protocell is formed was truly *random*, as orthodox Darwinism claims, it is reasonable to ask: What is the *probability* for such a random process to result in cells with *lefthanded* (1) amino acids in all twenty cases?

The answer is rather simple:

$$P(l) = (1/2)^{20} = 9.53 \times 10^{-7},$$

a very small probability.

The fact is, however, that *all life* in our planet, from the simplest protozoan to the bodies of the 7.0 billion men who populate today the Earth are made of *lefthanded* amino acids and *only* of *lefthanded* aminoacids.

"Explaining" the origin of life in terms of "*random biochemical reactions*" does not work. We, honestly, do not know what the reason for this unexpected non-randomness is. There may be a notable coincidence of natural factors, physical and chemical, which favour protocells with lefthanded amino acids. But like many other natural occurrences the origin of life does not look at all *chaotic* (random).

Orthodox neo-Darwinism, built upon the hypothesis that early life, and the whole tree of life, developed, quite undirected through (random) mutations and (random) natural selection is presented today as the explanation, and the *only* acceptable "*scientific*" explanation of life in our planet.

Simply, this is not true: for the moment and perhaps for

ever there is no such rigorous *"scientific"* explanation. It is true; however, that where there was no life billions of years ago there is life today. And that where there were only primitive living being many millions of years ago there are fully developed vertebrates today. But to say that all this was due to an *undirected*, chaotic and random evolutionary process is an unwarranted assertion.

Certainly, modern *biology* and *biochemistry* are much more than what is asserted in the statement "survival of the fittest" (a statement borrowed from Malthus by Darwin). In most cases it is not obvious "who" is the fittest. If it is meant that the *fittest* are "the ones who survive", and that the *"surviving ones"* are the "fittest", that does not qualify as a "scientific" statement. Modern biology is, of course, much more than that, but it is far from providing a complete "scientific" explanation of the origin life, at least for the time being.

Darwin's (and *Wallace*'s) contributions to modern biology were no doubt very important. But the contributions of *Schleiden* and *Schwann* (the cell), *Pasteur* (vaccinations, fermentation chemistry, optical activity, ...), *Mendel* (the laws of genetics), *Cajal* (the nervous system), *Von Laue* and the *Braggs* (X-ray diffraction), *Watson* and *Crick* (the double helix structure of DNA), *Flemming* and *Chain* (penicillin) were at least as important.

Let us quote[3,4] two very distinguished biologists (both Nobel Prize winners) on the origin of life on Earth:

"The modern cell's translating machinery consists of at least fifty macro-molecular components which are themselves coded in the DNA: the code cannot be translated except by the product of the translation. It is the modern expression of "omne vivum ex ovo"...

(Jacques Monod)

"An honest man armed with the knowledge available to us now could only state that in some sense, the origin of life appears at the moment to be almost a miracle, so many are the conditions which should have to be satisfied to get it going."

(Francis Crick)

It is certainly incredible that purely random processes were able to construct functional proteins or genes whose complexity is utterly beyond the creative capabilities of man's intellect.

Consider the *bacterial flagellum*. According to *Thomas Wood*, its functioning requires forty constituent proteins. The probability of such a random coincidence as that needed to produce a functioning flagellum would be of the order of

$$P(\text{flagellum}) = 10^{-1170}$$

Since the number of genes in the DNA of a higher animal is more than one hundred times larger than that of a simple bacteria, the probability of an actual random coincidence as that needed to produce a fully specified higher animal would be

$$P(\text{higher animal}) = 10^{-10000}$$

The purely combinatorial probability of a specific sequence of all atoms in the entire universe (10^{80}) is only a very small fraction of P(higher animal).

It may be pointed out, on the other hand, that Charles Darwin, as noted by Silvano Borruso[6], introduced significant changes in successive new editions of "The Origin of the Species". In the first edition he had said that "natural selection" could have provided a particular race of *bears* with characteristics better and better adapted to the aquatic medium so as to end up with a creature "as monstrous as a *whale*". This observation was suppressed in subsequent editions.

Man's intellect does not have many options confronting reality:

Subjectivism: Everything is a product of my own intellect, resulting in "solipsism".

Naturalism: Everything (including the subject) is part of a single existing entity, resulting in "materialistic monism".

Realism: There are "*intelligent*" subjects and "*intelligible*" (as well as "unintelligible") objects, both freely created and freely kept in existence by a Creator who is the only *"Unum Necessarium"*.

Science deals with observable, measurable phenomena in nature involving matter and energy under the action of *secondary causes*. This does not exclude, rather, it requires a *Primary Cause* beyond those secondary causes.

When Saint Paul wrote to the Romans:

"Ever since the creation of the world (God's) invisible nature namely, his eternal power and deity, has been clearly visible in the things that have been made... Men have no excuse for their foolish behaviour".

(Rom.1:20)

The creation of space-time, of matter and energy, in a primordial event at which the *thermal history* of the *universe* begun; the marvellous conditioning of our *galaxy*, our *solar system* and our *Earth* as a truly privileged planet; the origin of *life*, the development of the tree of life; life's stability in this privileged planet; and finally the creation of *man*, endowed with intellect, will and freedom, are not unintended *random occurrences*.

The *natural order* does not exclude a *supernatural order*. Rather, it *requires* one. But, also, and this is very important, it must be recognized that the *"Unum Necessarium"*, the Primary Cause, the Intelligent Creator was free to create or not . Otherwise the Creator would become a part of nature and would not be free becoming and element of a pantheistic nature.

The need of a Creator

The universe, earth, life, man are neither purely *random* (chaotic) occurrences, nor strictly *necessary* (deterministic) brute facts: they are contingent, i.e. *freely created*, freely made by an Intelligent Creator, that is precisely why they are intelligible realities, at least partly intelligible.

Jacques Monod[7] in "Chance and Necessity" says the following:

"We would like to think ourselves necessary, inevitable, ordained from all eternity. All religions, nearly all philosophies, and even a part of science testify to the unswearing, heroic effort of mankind desperately denying its own contingency".

He seems to use here the term "contingency" as interchangeable with "randomness". But, in the classical sense (in which it was used by the great masters of the 13th century), to be "contingent" is to be "not necessary", i.e. "created", *"freely created"*.

Benedict XVI, on the other hand, said in his inaugural papal mass[8]:

"We are not some casual and meaningless product of evolution. Each of us is necessary".

Here the term "necessary" must mean "willingly created" by God, i.e *"freely created"*.

As noted by *Anne Barbeau Gardiner*[9] (NOR, March 2011) commenting upon "God and Evolution":

"Here is what we're up against today: Two out of three college biology teachers call themselves atheists or agnostics, as do ninety five percent of biologists in the National Academy of Sciences. Of the leading scientists involved in evolution, eighty seven percent deny the existence of God, and ninety percent reject any purpose in evolution, "the creation story" for atheists, now operates "as the normal stance of sciences". In high-school and college textbooks, Darwinist evolution is taught as a a blind, heartless, purposeless, unguided process that makes any spiritual explanation of life superfluous. This is our current tax-funded orthodoxy enforced by court orders. Worst of all, what is "almost universally taught in textbooks" is that man himself is the unintended byproduct of blind material forces..."

If the Creator is omnipotent, everything which takes place does so because He wills it or because He allows it. A Creator which is not omnipotent is not a true Creator.

All the world civilizations, some defunct, some very much alive, are merging today in the present *global civilization* which, with roots in the Catholic Medieval European civilization, expanded itself rapidly throughout the nineteenth and twentieth centuries on the wings of modern *science* and *technology.*

Those scientific developments, which gained impetus in the century previous to the last World War resulted after 1945 in a tremendous, polifacetic , *technological revolution* which gave rise to a fourthfold increase of the world population, from about *1.5 billion* at the beginning of last century to *6.0 billion* by the year 2000.

But today, specially in the most developed countries, modern man is increasingly disgusted with the world, whith himself and with everything under the Sun. He has lost a proper perspective to look at temporal realities (fugitive realities) because he has lost the ability to appreciate eternal realities.

The *"cultural matrix"* producing this profound sickness could be called, for lack of a better name, *Evolutionary Materialism.*

There is a common denominator which underlies *Marx*'s communism, *Darwin*'s blind random evolutionism, *Nietzsche*'s moral nihilism. It consists in the reduction of man to the category of "primate", a "primate" competing with other predators for food, pleasure and power.

But it is undeniable that material goods alone, even if plentiful, cannot fully satisfy man's higher aspirations. Modern science tells us that the physical universe in which we all live is *finite*: Finite in time, finite in extension, finite in mass and energy. After *13.7 billion* years of cosmic expansion; after *4.8 billion* years of planetary existence, after *15 thousand years* of civilized human existence; after only 3 centuries of scientific and technological development mankind today is at a *crossroad,* It confronts its own *moral* and *social* decomposition in the middle of an unprecedented material wellbeing brought up by scientific and technologic progress.

Modern man finds increasingly difficult to cope with natural catastrophes as well as with manmade catastrophes. Suffering has become something intolerable for him. He has lost the supernatural perspective to look at it.

On the other hand, the "plate tectonics", which is the origin of all earthquakes and "tsunamis" producing today so many victims, is the same natural phenomenon which did contribute decisively, billions of years ago, to the formation of an atmosphere friendly to life on the earth's surface.

Modern medicine has been impressively successful to check many human plagues (tuberculosis, cholera ...) but is not omnipotent.

The *competition* between individuals and species is part of the *natural order*. Something which benefits one particular species or one particular individual may be detrimental for another species or another individual. Death and suffering are natural phenomena.

All of this takes place, of course, in a *definite time scale*. The lifespan of some insects may be days, while the lifespan of elephants is one hundred years, one hundred thousand times longer. The lifespan of man in this world is not meant to be for ever.

With so much in common with the higher animals, men are obviously quite different from the rest. According to the *Biblical account* man (distinctly, man and woman) was created in God's image and likeness. He was endowed with intelligence, will and freedom, within the same ecosystem made up by plants and animals. But he is called to go much higher.

Man is endowed with an *immortal soul* which sets him immensely *above* all those higher animals. He is a religious animal, and science is not a substitute for religion.

As noted by Sir William Bragg[10], a very distinguished Physics Nobel laureate, and a good Christian:

"From religion comes Man's purpose; from science his power to

achieve it. Sometimes people ask if religion and science are not opposed one to another, It is an opposition by means of which anything can be grasped"

And, as G. K. Chesterton said, for those who have chosen to believe instead of to doubt:

"It may be accepting a miracle to believe in free will, but it is accepting madness, sooner or later, to disbelieve in it".

REREFENCES

[1]Michael Danton, "Nature's Destiny" (The Free Press: New York, 1998).

[2]Manuel Mª Carreira, SJ, and Julio A. Gonzalo. "En torno al Darwinismo" (Ciencia y Cultura: Madrid , 2009)

[3]Jacques Monod, "Chance and Necessity" (Collins: London, 1972).

[4]Francis Crick, "Life Itself" (Simon & Schuster: New York 1981).

[5]Thomas Woodward, "Darwin strikes back" (Baker Books: Grand Rapids, Michigan 2006).

[6]Silvano Borruso, "El evolucionismo en apuros" (Criteria Libros: Madrid, 2001).

[7]Monod, Ibidem.

[8]Inaugural Papal Mass (April 24, 2005).

[9]Anne Barbeau Gardiner (New Oxford Review, March 2011)

[10]Sir William Bragg

6. The purposeful animal
by Manuel M. Carreira, SJ.

Within the marvelous richness of life on Earth we encounter as something very special the presence of Man, part of the animal kingdom, but exhibiting a peculiar type of activity that places human life in a new and superior level.

Living things have patterns of activity, due to genetic programming, that do not require learning or presuppose a conscious choice. Those instincts are so necessary for the survival of each individual and of the species that they practically define animal life, since they determine the basic functions of nourishment, self-protection and reproduction.

Even at the non-human level, evolution seems to show that living matter has a tendency to develop new forms to the limits allowed by the physical forces that determine biology, while this tendency appears to depend upon random factors, without finality. We face a contradiction when discussing life and evolution: *chance cannot be the sufficient reason for any kind of order or purposeful activity, and this kind of activity is the most obvious property of living things.* And purpose is certainly essential to understand the nature and free activity of Man.

Intelligence and consciousness

The specific activity that sets humanity apart in the animal kingdom is *reasoning*, using abstract concepts. Not a type of fixed behavior, but an appreciation of ideas, which are not sense reactions to a concrete material object. Even ideas about matter are processed into universal concepts: this is how science develops, formulating statements that claim universal validity even for things that the senses cannot detect.

Ideas lead to immediate or ultimate causes, *to sufficient reasons, to finality or ethical and aesthetic values*: something not found in other levels of the animal kingdom. In this activity we include Philosophy, pure Mathematics, Poetry, true "culture": a way of interpreting our existence in the Universe, expressed in art, developing aims that give sense to our life, and to social structures. Culture is shared through a spoken or written language, reinforced by expressive activities, so that each generation benefits from the achievements of its forebears in a kind of learning that is not found in non-human animal life, and that furthers the purpose of human development.

The steps "ideas-meaning-consciousness" refer to aspects of a single process that constitutes the reality of rational life.

There is an **I**, a personal subject that unifies the experiences of sense perception, draws from them their common elements, synthesizes concepts, establishes their value, *chooses the means* to communicate their meaning, and rejoices with their qualities of order and harmony. The subject recognizes its own identity as the center of an independent, autonomous and purposeful activity, freely developed, even to the point of leading to a behavior that goes against the most basic instincts: we can recall countless people sacrificing their lives for a religious belief, for patriotic duty.

What we think explains the way we act, because it leads to value judgments concerning ethics, and to affective reactions: *we seek what we love and perceive as good*, even when that goodness is different from any reaction of the senses. God as the Supreme Good has led many people to the highest degree of love and selfless sacrifice, while God is known as totally different from anything we can imagine. The search for Truth, Beauty and Good, summarizes the entire *purposeful* activity of Man as a rational being, who tends towards a reality not found in any description of matter, that cannot be detected by any laboratory instrument, and cannot be attributed to any of the physical forces.

Consciousness is the primary datum of rational life: we have a reflex knowledge of the fact that we are *knowing* and of states and decisions that flow from what we know. Only in the most imperfect way do we know the brain itself: its tissues, neurons, signal processing, need to be studied using the same instruments and methods required to study the brains of other animals. We know the matter of our body as the source of sense activity, but the external stimuli are much more evident than the processes that occur in the sense organ itself. *Matter is not conscious of itself*: we still do not know how the excitation of the neurons can be related to the content of conscious thought, nor the way biological and psychological levels influence each other.

Every property or process suitable for an experimental check results in a quantitative measurement with the proper instrument, but a number is never adequate to describe consciousness, or the value of an idea, or its ethical or artistic connotations. We can verify that thought is accompanied by minute currents in the brain, but nothing we can measure indicates the truth or beauty of an idea. Each neuron behaves like a transistor in a computer, whose task is to pass or block a signal, but never to determine its meaning. Consciousness cannot be explained by the totality of electrical currents in billions of neurons: if each signal or cell has zero consciousness, their ensemble cannot have it either. Thought is not a secretion of the brain, and those who speak in such terms are assuming a philosophical prejudice that begins with the statement that the only thing that exists is matter, understood in an arbitrarily undefined way.

The origin of intelligence

Since *none of the forces attributed to matter* by the physical sciences appears as a sufficient reason for consciousness and intelligence, it becomes necessary to accept a different cause. A new non-material (spiritual) reality must be present in Man, intimately joined to the biological element, and making a whole capable of two different kinds of activity, with mutual influences but with diverse results. This is not a "dualism" that postulates two independent beings joined in a temporary union; but rather the acceptance of two real components that cooperate and influence each other as parts of *a single substantial being* that is meant to exist as such.

This union is difficult to understand, and we cannot clearly explain the mutual conditioning nor the fact that the person is one and remains as one through all the changes that the body undergoes through life, but the fact that we cannot explain it does not invalidate the reasons presented for its acceptance. The alternative would be to postulate an unknown and undetectable "force" to produce consciousness, a force already present in elementary particles, that would increase its efficiency as the material structure grows in complexity, not just in the number of particles. Such hypothesis changes de definition of matter, pushing it beyond the limits that are acceptable in the physical sciences, based upon experimental checks.

This is, nevertheless, the implication of those viewpoints that accept biological evolution as the only reason for intelligence, considered as the necessary outcome of a *greater brain development*, which is attributed to secondary factors. But even within the human race, intelligence cannot be correlated with brain size: Neanderthal Man had greater brain mass than we do, and the trend in the last 10 thousand years seems to be to some reduction of brain volume. In our time, people with brain tissue limited by hydrocephalic conditions, have shown no loss of intelligence, and in some cases have had an IQ quite above the average.

Materialistic evolutionism has included the suggestion that there was an initial genetic programming of organisms to have them develop intelligence, with the programming attributed to chance (or even to extraterrestrial visitors.) This hypothesis cannot explain the facts: a genetic code can only determine new organic structures or instinctive behavior, but not mental processes that have no material output. We cannot program a computer to be self-conscious or to act freely to choose what to do and why, even if it is possible to have its functions determined by chance numbers as a response to data received from the human operator or from some instrument.

It is still possible to insist upon the trite examples of blind processes (either deterministic or due to chance, neither of them related to purpose) achieving the same results as human intelligence for the production of literary or other artistic works. The permutation of a few letters will lead to meaningful words being written automatically, and the systematic rearranging of billions of symbols of an alphabet will lead to the production of all the possible literary works that can be written with those symbols. This cannot mean that results attributed only to intelligence and conscious purpose can be obtained blindly, or that such processes lead to what we call human intelligence. These examples are misleading, because they presuppose that the letters of an alphabet, and the words composed with them, have a meaning by themselves, when it is clear that the opposite is true: they are *arbitrary symbols* that required a previous and conscious determination of a relationship between shapes and meaning. We need a language, a writing system, a grammar, and those elements have to be known to the observer who checks the outcome of those permutations. Otherwise, we shall simply have stains on pieces of paper.

"Chance" is not a physical force, but only a word to indicate that we are talking about objects or events that are unrelated by any common cause. Thus chance is never a sufficient explanation for anything, and it cannot be said to be the reason for order, constancy or structure. To attribute to chance the reality of abstract human knowledge, where the highest degree of complexity and order is found (the reason for science, art, philosophy) is truly to dismiss as meaningless our rationality and end up saying that our human culture can be explained with a childish "just because".

To hope that some future theoretical progress will lead to the explanation of consciousness and intelligence, in terms of the forces and particle structures at deep levels of matter, hides a prejudice that denies the very methodology that it claims as its justification: a scientific status is postulated for something that has *no experimental proof in any known fact*, nor can be shown to follow from any well established theory. It belongs to the realm of science fiction.

We might reflect upon the fact that the "information age", where the most vital and fast growing technology appears, deals with something *that is in itself intangible and without physical properties*. We cannot equate a given number of different atoms —or electrical impulses in the brain- with a person's identity, without considering as of any value the ideas, the driving purposes, the scientific, artistic or ethical achievements, or the human experiences of the person.

Physical laws and human free will

An objection is frequently made to the acceptance of human freedom –and thus also of purposeful behavior- by stating that it is incompatible with science. From a deterministic viewpoint, science requires certainty in its predictions, at least in principle, and free acts cannot be predicted. Starting from an opposite postulate, it is said that any activity is a *chance event* observed within an infinite number of probabilities that *must be realized* for every set of initial conditions. In both cases the human certainty that we are purposeful and responsible for our actions, with ethical and juridical consequences, is dismissed as an illusion.

The probabilistic interpretation of Quantum Mechanics is taken to absurd limits by asserting that anything that is mathematically possible (it has non-zero probability) *must occur* as a real outcome of any process. This means an infinity of universes, postulated to accommodate all possible values of the wave functions that describe either microscopic or large scale systems. The simple fact that such ensemble of universes is gratuitously affirmed, without any possibility of verification, is enough to relegate the theory to a non-scientific "mythology".

The conflict between freedom and scientific predictions rests upon the idea that material laws have to explain something that is not due to matter. Physical laws allow me to describe in detail how I flex my arm, how the muscles, tendons and bones interact, to make the arm bend. But they cannot explain why the arm bends when I WANT. This is the core of the question: the physical explanation does not cover everything, just as the reflection of light from the pages of a book says nothing about the joy of reading a poem or the insight of understanding a mathematical formula.

Chaos Theory tells us about the limited certainty of predictions in complex systems. Their sensitivity to minute changes in initial conditions makes the future state unknowable in the long term. But science is not rendered impossible by such limitations, just as it is not destroyed by the theoretical impossibility of knowing what is happening within a black hole.

Free will is the basis for our responsibility, without which human society cannot exist, nor can there be a meaningful concept of duty, human rights and justice, of personal ideals, of purpose. It is obvious that no proponent of its denial wants, in real life, to accept its consequences. The final attitude is an absurd schizophrenia, contradicting with personal behavior and the demands made from others what was dogmatically presented at the theoretical level.

Humans are subjects of duties, leading to rights that guarantee their development as rational animals: rights to be cared for in a family, in order to be fed, to grow, to be educated. Rights also to engage in legitimate activities as an individual and in a society context. And because all human individuals share the same dignity in this regard, it is never permissible to degrade a person to the level of something merely useful for the whims or profit of others.

Slavery, abortion, euthanasia, genetic manipulation, deny the dignity that each human being receives -not from society or any kind of democratic vote- but from the very nature of being human, as a patrimony that each individual can never be deprived of.

The image of God

Christian revelation, and its biblical forerunner in the book of Genesis, provides us with a surprising "definition" of Man. Not by philosophically stressing his nature as a "rational animal", but by referring him to his Creator: he is "the image and likeness" of an Omnipotent Being, Eternal, All-knowing and infinitely Holy. In some way, he reflects the One who, by His very essence, is Absolute Perfection.

This is diametrically opposed to the way of thinking found in the mythologies of all other cultures where the gods are described as "images of Man", enlarged to a super-human scale, but with the features, passions, and even cruelties, found in human beings. Since mythologies are the product of the inventiveness of poets who develop ideas found in their societies, it is logical to expect that they will only extrapolate to those divinities their own human experiences and thoughts.

The biblical and Christian God is not so limited: time after time, His Being and His way of acting is stated to be "not like that of Men." His eternal existence places Him outside of time itself ("for You, a thousand years are like yesterday, already gone"); His Omnipotence, without being arbitrary, implies a total dominion over all reality, spiritual, material or historical, allowing the creation of something from nothing, and the ordering of human destinies. His Wisdom knows no limit, showing itself in the marvelous order of creation and in the incomprehensible depth of His providence. His Holiness, places Him on a new level of fidelity, of impartial justice and undreamed mercy: He is the ultimate source of all good. His very nature, different from that of the material world, transcends any spatial measure: the heavens cannot contain Him.

Faced with this overpowering description of a reality we cannot comprehend or express in our language, that surpasses all our imaginative efforts and the inventions of all cultures, it is surprising that Man is called His "image and likeness", as the Bible puts it from the first moment when it speaks of human existence. The reason cannot be sought in some kind of corporal shape or in any other "likeness" due to superficial or changing qualities. It must be a likeness based upon the most intimate level of the human person, which in its powers and activities must reflect the way God is and acts. And God acts intelligently and freely, for a purpose.

This is shown when Man is presented with authority and dominion over the Earth and all its living creatures; he exercises this power by assigning to each a name. *He is intelligent, and he can act freely;* he is made from clay, but he has a "breath of life" that comes from God Himself, and he is destined to live forever, as God does. As the culmination of the creative process, he is *the purpose for which everything material is also created.*

When sin destroys the original closeness between God and Man, its immediate consequence is the loss of the immunity from death: Man ceases to be like his Creator, whose essence is to live forever. To be "an Image of God" is the glory of Man, as the most perfect of all creatures of the material world; "To be like God" is the temptation to seek a self-sufficient and absurd independence from the Creator. The distinction is crucial: the purpose of the second attitude is to avoid God, whom Man tries to challenge as a rival; the first title implies an affectionate nearness, because the best image -real and alive- of any living being, is a Son. And God is essential Life.

When, in the New Testament, God gives us the final and clear revelation of His nature, He reveals Himself as a Trinity, a family. It is the Son, the Image of the Father, who as the "Word" expresses all that the Father is in so far as His Being can be known by us. Christ is "the Image of His Substance", the perfect and essential Image because of His divine nature, and also an Image by being a Man, perfectly fulfilling the likeness that the Creator sought in Adam and Eve.

The early books of the Bible show no explicit hope of anything beyond earthly life: the just man is rewarded with wealth and children to carry his name, but nobody can praise God after death and human life appears as ultimately pointless. Only in later times, in the book of Maccabees, do we find a belief in life after death, a life that implies a new body given by the Creator to those who have died for their fidelity to His Law. We have to look for the outcome of this long process in the beliefs of sincere Jews at the time of Christ: Christ speaks about the resurrection assuming these beliefs, without causing surprise among His listeners.

Greek Philosophy was very soon put to use in Christianity to express with greater clarity the reality of a human nature where matter and spirit, body and soul, form only one person, which needs to be constituted by both elements in order to be truly human. In Greek thought we do not find any disregard for matter as evil, nor is Greek culture "materialistic" either: the spirit of Man is paramount, with the highest value placed upon wisdom, art, and freedom. Not surprisingly, these cultural viewpoints were adopted as being well suited to express the Christian belief regarding our nature and our future.

Because Man is an Image of God, we have to be an assertion of life with our entire being, not only with our spirit. The scientific description of the evolving Universe leads to the prediction of future conditions that will be incompatible with life of any kind, based on matter. But the presence of a non-material reality in Man, of a spirit, which cannot be explained by any evolution due to physical laws, leaves the door open to a survival, at least, of that new element. To that basic possibility, Christian faith adds the promise of a full human existence that confers immortality and eternity -a life of being "like God"- even to our material bodies.

An existence that imitates the eternal life of God cannot be subjected to the flux of time or the need for food and air: we shall be, in the words of Christ, "like angels". Even our bodies will be "spiritual bodies", free from the restraints of physical laws: they will be images of the risen Body of Christ. Thus the Son, Image of the Father, will be reflected in each human being, called to partake of the Life of the Trinity. The very temptation of Genesis will become a promise with a new and fuller meaning: to be "like God" will no longer imply a rebellion, but a fulfillment of our very nature, created in the "Image of God" our Father, intelligent and free, living forever.

.

7. The thinking animal
by Manuel M. Carreira, SJ.

Within the extraordinary richness of life on Earth we encounter as something quite special the presence of Man, certainly a part of the animal kingdom, but exhibiting a new type of activity that places human life in a new and superior level. We must deal with the problem of explaining this new way of being and acting, to examine it within the evolutionary context and to establish the limits of biological explanations for the fact of intelligent life in all its complexity.

We must begin by stressing our links to other animal life forms: all life on Earth is, in a real sense, a single phenomenon with common properties. Perhaps other types of life did arise long ago that did not develop into forms lasting up to now: the fact is that all present life forms show such a set of common parameters and properties that it becomes necessary to accept for them a single root. We utilize the same aminoacids, the same molecules with the same left-handed symmetry, the same DNA molecule for genetic coding. The water and carbon-based metabolism, with oxidation processes, is also the same in all living species, no matter how different their environments might be or their body plans. And all living things have innate patterns of activity, acquired by genetic programming, that do not require learning, nor do they presuppose consciousness. These instincts are so necessary for the survival of each individual and of the species that they can be used to define animal life, since they determine the basic functions of nourishment, self-protection and reproduction.

For any kind of animal activity the interaction with the environment is a necessary prerequisite: this is the role of the senses, material structures that react to some kind of physical input. Reactions are found, in almost every case, to temperature, pressure (touch), chemical agents (taste and smell) and to sound and light. All this is applicable to Man as well, with differences in degree with respect to other animals, just as there are differences among diverse species. Even if there are sense organs that we do not possess (for example, the electric sense that allows some fishes to detect their prey or some obstacle by the distortion it causes in the electric field originated by the animal) their nature is not totally different, since they react to the same forces -interactions- of matter that Physics describes.

Man is, structurally, a vertebrate, with a nervous system centered in the brain and the spinal cord, and with the same basic organs that we already find in fish for digestion, blood flow, motion and reproduction. The resemblance becomes quite remarkable when we compare human anatomy to that of terrestrial mammals, and finally to that of living primates: human genetic material is 98% identical to that of the gorilla. Logically, if we accept evolution as a fact at all previous levels, we must also do so when we look at the human body: it did not arise independently, as something totally new, but rather as related to the entire plan and history of life on the planet Earth, with an obvious conditioning by the unique events of its evolution during billions of years. This is so important that the history of life appears as impossible to duplicate: any change in concrete happenings, from the impact of a cosmic ray upon the nucleus of a cell to the catastrophe due to a giant meteorite, would lead the evolution of life to new and unforeseeable paths. Thus we cannot predict what evolution would take place in another environment, even in the case of another planet initially very similar to Earth.

We do find, nevertheless, that different evolutionary tracks eventually reached similar outcomes: the eye of an octopus is just as complex and efficient as that of mammals, and the same can be said about its nervous system. It seems that living matter has an innate tendency to develop new forms and abilities to the limit allowed by the physico-chemical forces that determine biological structures and activities, even if this tendency seems to be dependent upon factors that, considered in each concrete case, appear as random and without any finality. This is the puzzle or contradiction we face when discussing life and evolution: *chance cannot be the sufficient reason for any kind of order or purposeful activity, and this kind of activity is the most obvious property of living things.*

Intelligence and consciousness

Man is defined as a "rational animal", where the specific activity that sets humanity apart in the animal kingdom is *reasoning*: the use of abstract concepts. We are not concerned with a type of behavior, possibly learned by some kind of imitation, but rather with an appreciation of ideas, which are not just images or sense reactions to a concrete material object. Even ideas about something material are processed into a universal concept: I can think about *Carbon* as an element even if there is no atom of it present to me -and even if I have never found it in my experience- and the logical application of that concept must be correct always and in every place where Carbon is found. This is how science develops, studying the general cases beyond the data of individual experiments, and formulating its statements with claims of universal validity even for those things that the senses cannot detect.

The development of ideas, obtained by abstraction, inference or deduction, allows us to search for immediate or ultimate causes, for a sufficient reason, for finality or ethical or aesthetic values: in all cases things that cannot be perceived by the senses and that are not sought at other levels of the animal kingdom. In this activity we encounter Philosophy, pure Mathematics, Poetry, the most satisfying achievements of what is properly meant by "culture": *a way of interpreting our own real existence, and that of the Universe we are part of, that leads to artistic creations and to a set of ideas that give sense to our environment and to social and religious structures.* Culture is then shared through a process that utilizes arbitrary symbols of a spoken or written language, reinforced in many cases by expressive activities, so that each generation benefits from the achievements of its forebears in a kind of learning that is not found in any degree in non-human animal life.

The steps "ideas-meaning-consciousness" refer to different aspects of a single process that forms the reality of rational life. There is an "I", *a subject that brings together the multiple experiences of sense perception, draws from them their common elements, synthesizes concepts, establishes their value, chooses the means to communicate their meaning, and rejoices with their qualities of order and harmony.*

In all of this, the subject recognizes its own identity as the center for a true independent and autonomous activity, freely developed, even to the point of giving such value to an idea that a behavior can arise from it that goes against the most basic instincts: we can just recall how many people have sacrificed their lives for a religious belief, for the sake of honor or of patriotic duty.

What we know explains the way we act, because it leads to value judgments concerning ethics, and to affective reactions:

we seek what we love and perceive as good, even when that goodness is totally different from any pleasurable reaction of the senses. God as the Supreme Good has led many people to the highest degree of love and selfless sacrifice, even if religious faith presents God as totally different from anything we can perceive through our senses or that we can imagine. The search for Truth, Beauty and Good, summarizes the entire activity of Man as a rational being, who tends towards a level of reality that is not found in any description of matter, and that cannot be detected by any laboratory instrument, cannot be specified with any measurement and cannot be attributed to any of the physical forces.

Consciousness is the primary datum of rational life: we have a reflex knowledge of the fact that we actually know something, and of the states and decisions that flow from what we know. Only in the most imperfect way do we know our own biological structure and functions, even the brain itself: Its tissues, neurons, signal processing, are just now described by recent advances in physiology, using the same instruments and methods to explore the human brain that are needed to study the brains of other animals. We know the matter of our own body as the source of sense activity, but we find that the external stimuli are much more evident than the processes that occur in the sense organ itself. Matter is not conscious of itself in any of the organic processes, even in the functioning of the brain; we still do not know how the excitation of the neurons can be related to the content of conscious thought, or the way biology and psychology influence each other.

In Physics matter is DEFINED by its interactions (operational definition) since we only know what things are by what they do. The entire material reality (particles, energy, the physical vacuum, space and time) encompasses all and only those things that can have some interaction by at least one of the four forces that physics recognizes: gravitational, electromagnetic, strong nuclear and weak nuclear. Any other meaning that might be given to the word MATTER must be justified by a description of some activity that is experimentally verifiable: one cannot arbitrarily include in its concept other properties postulated *ad hoc* to solve problems in other areas, if we want to avoid an irrational twisting of scientific language.

Stressing this rigorous use of scientific concepts, we have to specify the kind of activity of each force and the limits that it entails as a possible explanation of reality. Gravity causes exclusively a collapse of matter into bodies of different sizes, and their motions in orbits that can, in principle, be exactly predicted. It can also give rise to gravitational waves, even if they have not been detected directly, but are inferred from the loss of energy in stellar systems.

Nuclear forces maintain the heavy particles of the nucleus in a very small volume (strong force) or change one particle into another (weak force) and have such limited range that they can only act within the limits of the nuclear radius or the size of a particle.

The electromagnetic force explains the attractions and repulsions that lead to the impression of hardness, elasticity and impenetrability of the material bodies of our experience. It is also responsible for the structure and chemical properties of crystals and biological entities. And it explains the existence and propagation of energy as waves traveling even through the physical vacuum, in the form of radio waves, visible light, x-rays and gamma rays; also the fact that there are electric and magnetic fields which modify the properties of space and determine the trajectory of particles endowed with electric charge. It is, by far, the force with the greatest variety of effects and the one that is responsible for vital functions at the sense level, as well as for the structure of DNA and living tissue.

But this description makes it clear that *none of the effects we have listed includes or brings about consciousness, meaning or abstract thought*. Every property or process mentioned so far allows for experimental checks and quantitative measurements with the proper instruments, but a number is never the adequate description of consciousness or of the value of an idea or of its ethical or artistic connotations. We can verify that intellectual activity is accompanied by minute currents in the brain, but nothing we can measure indicates the truth or beauty of an idea. Each neuron behaves like a transistor in a computer, whose only task is to pass or block a signal, but never to determine its meaning.

Consciousness cannot be explained by the totality of electrical currents even in billions of neurons and in their interconnections: if each signal or cell has zero consciousness, their ensemble cannot have it either. Human thought is not a secretion of the brain, and those who speak in such terms are doing so because of a philosophical -not scientific- prejudice that begins with the statement that the only thing that exists is matter, understood in an arbitrarily undefined way.

The well-known scientist Stephen Hawking, in his book *A Brief History of Time*, states that when a complete unification theory is obtained (the TOE or Theory of Everything) that describes the four forces in a single theoretical and mathematical framework, the theory must be capable, in principle, of predicting that I am reading his book. This is a totally gratuitous assertion: neither the author nor any scientist can establish a causal relationship between physical forces and human freedom and consciousness.

The opposite viewpoint is presented by Roger Penrose (coworker with Hawking for some important theorems concerning black holes) who in his book *The Emperor's New Mind* holds that computers cannot be said to have any degree of intelligence, nor will they have it in the future: since they are only ensembles of gates to pass or stop a current, no matter how many are built into a system, they cannot give rise to artificial intelligence in the strict meaning of the word, nor can they have consciousness or free will, that require true intelligence and abstract thought. The most capable computers still know nothing about what they are doing nor have any initiative to do it, and their achievements can be duplicated by mere mechanical systems, even if these are much slower.

The origin of intelligence

As a consequence of what has been explained so far we can state that, since none of the forces attributed to matter by the physical sciences appears as a sufficient reason for human consciousness and intelligence, it becomes necessary in our logical process to accept a different cause. And since the four forces DEFINE the meaning of matter, the new cause cannot be matter. A new non-material (spiritual) reality must be present in Man, intimately joined to the biological element, and making a whole capable of two different kinds of activity, *with mutual influences but with diverse results.* This is not to advocate a "dualism" in the sense of postulating two independent beings joined in an unnatural and temporary union; it is rather the acceptance of two real components that cooperate and influence each other as parts of a single substantial being that is meant to exist as such. It is true that this union is difficult to understand, and we cannot clearly explain the mutual conditioning or the fact that the person is one and remains as one through all the changes that the body undergoes through life, but the fact that we cannot explain it does not invalidate the reasons presented for its acceptance.

The only alternative to this description would be to gratuitously postulate a "fifth force" to produce consciousness already in the most elementary particles, a force that would increase its effects as the material structure grows in complexity, not just in the number of particles. Such hypothesis changes the definition of matter, pushing it beyond the limits that are acceptable in the physical sciences, based upon experimental checks.

This is, nevertheless, the implication of those viewpoints that accept biological evolution as the only reason for human intelligence, considered to be the necessary outcome of a greater brain development, which in turn is attributed to secondary factors like the bipedal stance of the primates, their phonetic ability or, simply, the survival value of intelligence.

It is certainly true that our survival is helped by our being intelligent, but this is due, in great measure, to the fact that as a simple biological entity, the human animal is totally incapable of self-protection during a very long infantile stage of development, and it is still very lacking in specialized instinctive skills in the adult period of life. If intelligence were to be developed gradually to overcome such limitations, primitive Man would not have survived: intelligence had to be found in full from the first moment that human beings needed it.

We can also point out the weakness of the explanation when applied to the survival, during hundreds of millions of years, of many species that exhibit zero intelligence (in the strict sense we have described): insects and dinosaurs remained and flourished on our planet during periods greatly exceeding the existence of the human species. The same is true of the simple micro-organisms that are still, by far, the most abundant -most successful- type of life on Earth.

The causes presented as leading to the development of intelligence, due to organic changes, cannot be taken as a sufficient reason either. Even if the upright stance frees the brain from a compression against the spine and allows for its development, the cause-effect relationship is dependent upon the assumption that more neurons in a more complex pattern necessarily lead to intelligence, just as if thought were a secretion of the brain. And the evolutionary path that led other animals to forms that would seem likely to result in a similar development did not, in fact, produce similar results. The ostrich could have developed a large brain, but it did not; the dolphin and the elephant have larger brains than we do, but do not have intelligence. Even within the human race, intelligence cannot be correlated with brain size: Neanderthal Man had greater brain mass than we do, and the trend in the last 10 thousand years seems to be to some reduction of brain volume. In our own age, people with much reduced brain tissue, due to hydrocephalic conditions, have shown -in general- no reduction of intelligence, and in some cases have had an IQ quite above average.

Still less convincing is the logical inversion of attributing the development of intelligence to the capability of producing well modulated sounds, implying that the possibility of developing a language led to having something to say: this is clearly not applicable to a parrot, who can be trained to "speak" in several languages, but has no ideas of its own to express. To assert that a syntactic structure is present in the brain, previously to any formal language and to its conceptual content, is to invert the logical order. We choose the way to express our meaning according to the ideas we want to communicate, and there is no sound or visual symbol that obliges us to have a particular set of ideas. On the contrary, we can express the same ideas in several languages with totally different grammatical and syntactic rules, even if a given language might seem better adapted to present some concrete scientific, philosophical or poetic insight.

The "emergence" of intelligence from matter can only mean that it was already present, in some hidden form, in the previous stages of material structuring. This remains an unproven assertion, without any scientific or philosophical basis, and it is incompatible with the principle of sufficient reason if we start with the accepted definition of matter. It simply introduces a new word to say, in a disguised form, that the only thing that can exist is matter and that anything that happens has to be attributed to it. This kind of materialistic reductionism was already proposed by ancient philosophers, was taken as a central dogma by Marxism, and it has been proposed repeatedly by positivist thinkers in our time. But changing the words does not make it more believable, because it does not add anything new as a real explanation; its acceptance has to be due to some prejudice or to the uncritical following of some system or author proposing it.

Some forms of "emergentism" have included the suggestion that there was an initial programming of living organisms to have them develop intelligence, with the programming attributed perhaps to extraterrestrial visitors.

But even this far-fetched hypothesis cannot explain the facts: any kind of genetic code can only determine new organic structures or instinctive behavior, *but no ideas or mental processes that have no material output*. We cannot program a computer to be self-conscious or to act *freely* to choose what to do and why, even if it is possible to have its functions determined by chance numbers as a response to data received from the human operator or from some instrument. Or the program can be set to attain an end that *the programmer* –not the machine- defines as desirable.

It is still possible to insist upon the trite examples of blind processes (either deterministic or due to chance) achieving the same results as human intelligence for the production of literary or other artistic works. The permutation of a few letters will lead to meaningful words being written automatically, and the systematic rearranging of billions of symbols of an alphabet will eventually lead to the production of all the possible literary works that can be written with those symbols. This is taken to mean that results that are attributed only to intelligence can be obtained without it, and that similar processes over a long period of time do lead to what we now call human intelligence. But these examples are really misleading, because they presuppose that the letters of an alphabet and the words composed with them *have in themselves a meaning*, when it is clear that the opposite is true: they are arbitrary symbols that required a previous and conscious determination of a relationship between shapes and meaning. *There has to be a language, a writing system, a grammar, and all those elements have to be known to the observer who checks the outcome of those permutations.* Otherwise, we shall simply have stains on pieces of paper.

Another strange example of matter producing meaning proposes the reaction of a primitive human to modern technology as a basis to deny the logic of seeking non-materialistic explanations. It is said that a jungle dweller, out of touch with our modern developments, would automatically attribute to some "spirit" the images and stories displayed in a

TV set, while we know that the only thing in the set is an electrical impulse that makes the screen glow here and there. Then it is said that we tend to look for a "spirit" in our brain to explain thought, but that only electrical currents are needed.

The example is just a proof of the opposite view: *the TV show is not produced by the electrons*, but by a human mind responsible for the plot and the way to present it. *Nobody blames the electrons* for a boring show, and the simple electrical currents, when nothing is transmitted from the station, can only produce chance points of light here and there, without any informational content. There is a spirit behind the show, even if the receiver is far from the origin of it.

We should also stress that "chance" is not a physical agent or force, but simply a word to indicate that we are talking about objects or events that are truly unrelated by any common cause. Thus *chance is never a sufficient explanation for anything, and it cannot be said to be the reason for order, constancy or structure.* To attribute to chance the reality of abstract human knowledge, where the highest degree of complexity and order is found (the reason for science, art, philosophy) is truly to renounce our rationality and end up by saying that the marvelous achievements of human culture can be explained with a childish "just because".

The final refuge of materialistic reductionism is the *a priori* and blind conviction that some future theoretical progress will lead to the explanation of consciousness and intelligence in terms of the forces and particle structures at deep levels of matter. If one does not stick to scientific rigor, no proof will be sufficient to show that this will not happen. But this attitude hides a prejudice that denies the very methodology that it claims as its justification: a scientific status is postulated for something that has no experimental proof in any known fact, nor can be shown to follow from any well established theory. It belongs to the realm of science fiction.

Physical laws and human free will

An objection is frequently made to the acceptance of human freedom by stating that it is incompatible with science. From a deterministic viewpoint, science requires the possibility of certain predictions, at least in principle, and free acts cannot be predicted. From an opposite starting postulate, it is said that any activity is a chance event that is observed within an infinite number of probabilities that have to be realized for every set of initial conditions. In both cases the human certainty that we are responsible for our actions, with the consequent responsibility in ethics and before the law, is dismissed as an illusion.

The probabilistic interpretation of Quantum Mechanics is taken to its limit by asserting that anything that is mathematically possible (it has non-zero probability) *must occur* as a real outcome of any process. This means that an infinity of universes must be postulated to accommodate all possible values of the wave functions that describe either microscopic or large scale systems. The simple fact that such an ensemble of universes is gratuitously affirmed *without any possibility of experimental verification* is enough to relegate the theory to a non-scientific "mythology", which does not deserve other reply than the requirement of proofs for something that totally ignores the methodological rigor of "Occam's razor".

The conflict between freedom and certain predictions rests upon the idea that material laws have to explain something that is not due to matter. Physical laws allow me to describe in detail how I flex my arm, how the muscles, tendons and bones interact, with tensions and energy releases that finally make the arm bend. But they cannot explain why the arm bends when I WANT. This is the core of the question: the physical explanation does not cover everything, just as the physical description of my observing a beautiful sunset does not include my pleasure in seeing it, and the reflection of light from the pages of a book says nothing about the joy of reading a poem or the insight of understanding a mathematical formula. Einstein himself, when asked if he thought that some day everything could be expressed in physical terms, answered that such an effort would lead to the absurd of equating the beauty of a Beethoven sonata with the graph of air pressure variations when the orchestra performs the work. The reality of the world, and our experience of it, is much richer than that, and no single method of knowing it is enough to cover its variety.

We should also take into account what the modern Chaos Theory tells us regarding the certainty of predictions in complex systems. Their sensitivity to minute changes in initial conditions makes the future state unknowable in a sufficiently long term. We cannot predict the position of a planet in our solar system millions of years in advance, because we would need to establish the present position of all the bodies with a precision of millimeters, and their masses and velocities with comparable accuracy. But science is not rendered impossible by such limitations, just as it cannot be destroyed by the theoretical impossibility of knowing what is happening within a black hole.

Our free will is the basis for our responsibility, without which human society cannot exist, nor can there be a meaningful concept of duty, human rights and justice. It is obvious that no proponent of its denial wants, in real life, to accept its consequences. The final attitude is, then, that of an absurd schizophrenia, contradicting with personal behavior, and the demands made from others, the position that was dogmatically presented at the theoretical level.

Because humans are subjects of duties, they are also subjects of rights that guarantee their development as rational animals: rights to be cared for in a family in order to be fed, to grow, to be educated; rights also to engage in all legitimate activities as individuals and in a society context. Because all human individuals share the same dignity in this regard, it is never permissible to degrade a person to the level of a thing, something merely useful for the whims or profit of others. Slavery, abortion, euthanasia, genetic manipulation, deny the dignity that each human being receives -not from society or any kind of democratic vote- but from the very nature of being human, as a patrimony that each individual may never be deprived of.

Matter and spirit

The concept of a non-material reality as an explanation for the existence of human intelligence is equivalent to the acceptance of a soul, a spirit, that cannot be understood only in terms of physical parameters nor can be imagined, but that is inferred as the sufficient reason to explain consciousness, intelligence and free will. *Its existence cannot be justified by any material development tied to genetic programming*: not having a material structure, it cannot split nor reproduce itself using any "prime matter" in its surroundings.

It has to be attributed to a strict "creation" by a spiritual Being with infinite Power (required by any creation from nothing). This is applicable to every human individual as well as to the whole human race.

The moment and the historical context of the creation of humanity are unknown to us at present. It is possible to have long periods of human existence devoid of any archaeological remains to prove it: even now there are jungle dwellers who do not use stones, but only wood, and who do not leave any permanent indication of their activities, but who are just as intelligent as any other modern human. The first clear proof of humanity is found in the appearance of complex utensils, some times with elaborate decorative incisions and colors; cave walls with paintings of remarkable artistic quality, burials that imply a conviction of some type of survival after death.

This desire of survival after death is present in all cultures: in spite of the obvious fact of death and corruption, apparently identical to that of other animals, religious ideas are developed that in some way deny our own disappearance. The "I" who remains as the subject of so many different activities and changes during our life, is expected to exist still after the mysterious fact of our death, not only in the memory of friends or family, but in some kind of invisible world that, in many cases, is thought to be similar to the present world of our experience. Such survival is frequently considered dependent upon special rites or burial practices, even in our own age.

We might reflect upon the fact that, in our time, the production and transmission of ideas and data has given us the "information age", where the most vital and fast growing technology deals with something that is in itself *intangible and without physical properties*, and this is the best index of progress. But some materialistic writers think about our survival in such crass terms that they reach the point of equating our "immortality" with keeping in a laboratory the cancer cells that caused the death of a patient, or they suggest that immortality is achieved if our DNA program is kept in some kind of electronic memory. This is tantamount to equating a given number of magnetic domains with a person's identity, without considering as of any value the ideas, scientific or artistic, or the ethical achievements and human experiences of the person. This is a viewpoint more restricted and myopic than that of our forebears who thousands of years ago honored their dead with offerings in the tombs of the Stone Age.

Finally, we might ask about the future evolution of humanity. Nothing shows as an obvious bodily change in the history of Homo Sapiens, which only covers a rather short time within the span of evolution of living things. But because we are able to modify our environment, adaptation to it has ceased to be a pressing need, and this will be more and more evident as technology develops further. Our way of becoming more human has to be found in a greater development of that search for Truth, Beauty and Good that is our privilege and our duty, and in the application of those values to society at every level. When each person on Earth can truly be described as freely pursuing the ideal of perfection that befits our nature, the human species will have reached a peak that no biological evolution can ever lead to

.

RELIGIONS
OF THE WORLD

8. The great Religions of the world
by Julio A. Gonzalo

Man could be defined as the religious animal (or as the smiling, the artistic, the historian, the rational animal, for that matter).

From the earliest times Egipt (from 4.000 BC), Summer (from 3.000 BC), India (from 3.500 BC), China (from 1.500 BC) most men in any civilized society have had gods. Mosaic Judaism (since 1.300 BC), Greek polytheism (1.100 BC), Budism (483 BC), Confucianism (479 BC), Taoism (300 BC), Christianism (30 AC), Islam (622 AC), have offered prayer, sacrifice and homage to the gods or to the One God.

Catholicism, as the heir of Judaism and the troncal form of Christianism, which embraces Eastern Orthodoxy and Western Protestantism, has a unique feature which differentiates it from the other religions of the world: it summarizes the fundamental truths necessary for salvation in a condensed statement: the Creed of the Apostles. This Creed has been further explained and developed in the Nicene Creed, The Tridentine Creed and, most recently, in the Post Vatican II Creed of Paul VI. According to the Creed, Christ is the Son of God who became incarnated, was crucified and resurrected in the third day.

Coexisting from classical times with all this world religions there is another substitute religion, Militant Atheism, which since the French Revolution and, again, since the Russian Revolution, has become increasingly more influential. Russian Soviet Communism (and its heir, Chinese Maoism) for almost one century, has been an anti-religious and specially anti-Christian movement with one hundred million victims left for posterity, according to "The Black Book of Communism: Crimes. Terror. Repression" (Harvard University Press, 1999). It is appropriate to use "substitute religion" to characterize Soviet Communism because in its early phase it required mystic devotion and true spirit of sacrifice from its adherents.

Today's Western Militant Atheism is a relatively recent phenomenon which begun to become fashionable just as Soviet Communism fell down. Philosophically, it can be equated with radical "moral relativism" (divorce express, abortion, homosexual marriage, hostility towards the traditional family...). This relativism is the inevitable consequence of assuming evolutionary materialism as the starting point: if there is no such a thing as a human nature, man is an evolving animal, even more so that any other animal.

For Militant Atheism, truth and error, good and evil, beauty and ugliness are almost interchangeable concepts. Militant atheists do not like to be called "nihilists", but they are.

A distinct characteristic of Militant Atheism is its bold attempt to appropriate modern science for itself. In fact it is creating a whole new jargon of derogatory terms to condemn "a priori" no only anything Christian but also anything divine. This is done with undisguised arrogance, ignoring the fact that modern science, science proper, historically, has developed in one and only one civilization: Medieval Christendom, the cradle of post-Renaissance Europe.

If God created everything in time (cosmos, living beings and men endowed with intellect, will and freedom...) it is eminently reasonable to expect that He is not indifferent to his work. Man's present status is the status of a fallen creature, subject to error and evil, but God is always free to reveal himself to men. And revelation is given by God to men as a gift. And men, of course, have the God-given right to accept or to reject it.

God wills all men to love Him freely, and to obey Him freely, In principle, Jews, Christians and Muslims acknowledge the Ten Commandments, given to men in order to be happy in this life and in the life to come. All men, not only Jews, Christians and Muslims, have a God-given conscience to distinguish between good and evil. A conscience of which the Ten Commandments are just explicit reminders. That is why it is legitimate to talk about Natural Religion as distinct form Supernatural, Revealed Religion. God is always infinitely merciful, but also infinitely just. We are not entitled to tell him, like the farisee in the Gospel's parabola, that we are more worthy of his esteem that the ignorant publican praying at the back of the Sinagogue. The publican, however, pleased God more than the farisee because, recognizing his faults, he asked for God's forgiveness.

It is most natural in man to recognize that everything he has is received. In the natural order, the religions of the world, properly understood, teach men to be grateful to God and to men. This does not mean of course that all religions are equally true. Only a truly revealed religion would deserve an unconditional support.

James Clerk Maxell, one of the greatest physicists of all

times, a devout Christian, wrote to his friend Colin Mackenzie:

"Old Chap! I have read up many religions; there is nothing like the old thing after all", and continued, "I have looked into most published systems and I have seen that none will work without God".

("The life of J.C. Maxwell" by Campbell and Garnett 416, quoted in Karl A. Kneller, "Christianity and the leaders of Modern Science, "with an Introductory Essay by Stanley L. Jaki, Real-View. Books: Fraser, Michigan, 1995).

Let us review briefly the Great Religions of the World.

Hinduism. According to the Encyclopedia Britannica (Book of the Year 2013) 967 million men believe today in Hisduism (13.7% of world population). Hinduism is a complex religious system which in various regions has absorbed very diverse elements throughout many centuries. In the valley of the Indo river, a pre-Aryans people had a strong government and was well organized already in the 3rd millennium b.C. The Aryans invaded the valley from the North 1500 years b.C, and they brought their own sacred books written in Sanscript. As time went on the religion of the invaders assimilated the cults and rituals of the pre-Aryan people. This gave rise to modern Hinduism. A characteristic of hinduism is the system of castes, based in part in the segregation of common people by their work, and in part on the racial pride of the Arian invaders. There are in Hinduism three fundamental beliefs: the reincarnation (present life is only one of many lifes, and every man will have opportunity of a better life in the future; that is why living animals are in great esteem in India , they could have reincarnated persons) among hindus abortion has a very bad reputation. Previous existences determine his karma, and a man present position in life determines the dharma. his rights and duties in this life. A hindu must struggle to liberate himself from the ever recurrent cycles of successive reincarnations through meditation and yoga in order to join the Universal Being (Brahaman). Within a politeist wrapping, Hinduism is therefore a kind of Cosmic Pantheism, in which every thing is governed by a general Law of eternal returns.

Judaism. There are only 15 million Jews in the world today (see Encyclopedia Brittanica, Book of the Year 2013) but due to the paramount historical role played by the old religion of Israel, fist in relation to Christianity and also in relation to Islam, Judaism must be counted within the great religions of the world. Judaism, Christianity and Islam are called the "Religions of the Book" (the Bible), having in common Monotheism and the Ten Commandments among other things. The Pentateuch (Genesis, Exodus, Leviticus, Numbers and Duteronomy) gives God's revelation to Abraham, Isaac and Jacob, along with the promise of a Savior to be born of their descent. Moses is the great leader who brings the Jewish people out of serfdom in Egipt to the promised land. The prophets, which announce, many centuries in advance the coming of Christ, and the other great books written after the captivity of the Jews in Babylon, complete the Old Testament. After Christ's coming, the Old Testament is complemented by the New Testament, which includes the four Gospels, the Acts of the Apostles, the Epistles (specially Saint Paul's Epistles) and the Apocalipsis, announcing the Second coming of Christ, after the final great persecution.

God is holy and He demands his people to be holy, righteous and just. But his people becomes unfaithful again and again. At the time of Christ, the Sadducees did not believe in eternal life while the Pharisees did. The martyrdom of the Maccabee brothers would have no sense otherwise.

After their exile in Babylon, the Jews reconstructed the Temple of Jerusalem. But this second temple was destroyed by the Romans in A.D. 70 fulfilling Christ prophecy. The Catholic Church has claimed, from the 1st century, to be the true depositary of the Old Testament promises, and her liturgy reflects abundantly the continuity between the Old and the New Testament.

In the 20th century, after the bloody holocaust of millions of Jews during the Second World War, with the support of the United Nations, Jews from Europe and all over the world established in the Holy land the State of Israel. Since then Israel is in permanent conflict with the displaced Palestinian muslims.

Buddism. According the Encyclopedia Britannica (Book of the Year 2013) there are 504 million buddists in the world (7.2%). Buddism can be considered a heretic outgrowth of Hinduism which spread out to the East and became popular largely outside India, in Tibet, Indochina, China and Japan. Budda (in Sanskrit "the enlightened One") lived probably from 563 to 483 b.C. The earliest written accounts of his life are dated 200 years after his death. Siddartha Gautama was born the son of a king of the Sakya clan, belonging to the warrior caste. That is why he was called the Sakyamuni "the sage of the Sakyas" in the Himalayan foothills (South Nepal). It was predicted at his birth that he would be a world ruler, and teacher. His father, the king, took great pains to free him early from all misery and tried to avoid any religious influence upon him. So Siddartha spent his youth in great luxury. He married and had a son. At the age of 29 he left the palace grounds in his carriot to see more of the world. In successive trips he saw an old man, a sick man, a corpse, and a mendicant monk. He learned therefore that suffering and death are unavoidable, and he saw in the monk his destiny. Forsaking wife and son he became a wandering ascetic. After mastering yogic meditation from various teachers he undertook fasting and extreme austerities but, after six years, he gave up these austerities fearing they might bring death to him. On the night of the full moon, after overcoming the attacks of the "evil one", he reached enlightening at the age of 35. Then he proceeded to the Deer Park where he preached the first sermon to five disciples, "setting into motion the wheel of dharma", which contained the "four noble truths" and the "eight fold way" and spent the rest of his life as a teacher. Tradition says that he died at the age of 80. More than a religion, Buddism, therefore, seems to be a philosophy of life: If you want to avoid suffering, avoid to get involved, avoid to love, withdraw as much as possible within yourself. As G.K. Chesterton said:

"The Fall is a view of life. It is not only the only enlightening, but the only encouraging view of life. It holds, against the only real alternative philosophies, (those of the Buddist or the Pessimist or the Promethean) that we have misused a good world, and have not merely been entrapped into a bad one. It (the Fall) refers evil back to the wrong use of the will, and thus declares that it can eventually be righted by the right use of the will. Every other creed except that one (the Fall) is some form of surrender to fate."

("The Thing:Why I Am a Catholic", New York: Dodd, Mead and Co., 1946, p. 226)

Confucianism. The Encyclopedia Britannica gives 437 millions as the number of adherents to Chinese folk religions and only 8.2 millions as the number of adherents to Confucianism presently. Given the great role played by Confucianism in Chinese history, it may be assumed that Confucianism, mixed with Taoism and perhaps Buddism, is an important ingredient for folk religions. Confucius (Kung Fu-tse in Chinese) lived probably between 551 and 479 b.C. He was a Chinese sage who made many disciples. They collected his wise sayings and taught them. For instance: "Confucius said, He who exercises government by means of his virtue may be compared to the north polar star, which keeps its place while all other stars turn around it" ("Quotations of Confucius", Jinan, 2006 p.7). He was born in the feudal state of Lu, in the modern Shandong province. Upset by the constant warfare between Chinese states and by the tyranny of the rulers, he developed a system of morality and wisdom to foster peace and to give people a stable and just government. At the age of 55, for ten years, he toured several neighboring states to influence rulers, but he was not successful, probably due to his extremely outspoken manners. He is considered the author or editor of the Wu Ching (the five classics of Chinese literature). Confucianism envisaged man as essentially a social creature bound to his fellow men by Jen, a term meaning "sympathy" or "human-heartedness". This sympathy is expressed through five relations-soverain and subject, parent and child, elder and younger brother, husband and wife and friend and friend. Of these, the filial relation is stressed most. Etiquette and ritual are an important part of these sympathetic relationship. A person may be socially superior to some and personally inferior to others, but must treat properly both. Correct conduct proceeds not through compulsion but through a sense of virtue inculcated by subtle models of behavior. The ruler, as the moral exemplar of the whole state, must be irreproachable, but all men have also a strong obligation to be virtuous. The millennial "great commonwealth" united mankind under ethical rule would take a long time to achieve. In the 1st century A.D. offering sacrifices and paying veneration to Confucius developed in special shrines and it continued with varying intensity at various times and places up to the 20th century. The overthrow of the Chinese monarchy, with which Confucianism was closely identified, led to the disintegration of many Confucian institutions, accelerated after the victory of the Communist revolution (1949).

Taoism. Only 8.5 million adherents in the world are given in the Encyclopedia Britannica (Book of the Year 2013), but Taoism is likely to play an important role also among the 437 millions of Chinese folk religions believers. It is a Chinese religious and philosophical system derived from the book "tao-te-ching" traditionally attributed to Lao-Tze, (probably mid-third century b.C). The Tao means the way, in a broad sense, the path taken by natural events. It is characterized by spontaneous alternatives in natural phenomena taking place

REASON, SCIENCE & REVELATION

naturally with no effort, for example, day following light, or water going spontaneously by itself to the lowest level.

Man, following the Tao must free himself from all strings. His ideal state is freedom from desire. The political doctrines developed by the Taoists reflect a quietistic philosophy, in contrast with those of the Confucianists, which stress discipline and ordered behavior. Many features of Taoism were later adopted by the Mahayana Buddists. Taoism developed a large pantheon, incorporating many local gods. Monastic orders and Lay masters. Chinese literature, painting and calligraphy were greately influenced by Taoism. In the 1950's it was officially prohibited in China.

Christianism. According to the Encyclopedia Brittanica (Book of the Year 2013) there are 2.319 million Christians all over the world. 32.9% of the world population). Of them 1.200 are Roman Catholics, 845 are Protestants, Independent, Anglican or Dissident, and 277 are Orthodox, including Greek and Russian Orthodox.

Founded in Palestine in the 1st century by Jesus Christ on the foundation of the Jewish religion it spread out East and West soon throughout the Roman Empire. The central teachings of Christianism are: Jesus Christ is the Son of God, the second person of the Trinity (One God in three Persons: The Father, the Son and the Holy Spirit). Jesus Christ birth from a Virgin, announced many centuries before, his private and public life on earth, his crucifixion, resurrection and ascension to heavens, are proofs of God's love for men. All men, by their faith, accompanied of goods woks, are called to salvation and eternal life in heaven (Creed of the Apostles). The Old Testament (Jewish Bible) and the New (written by Christ disciples) complement each other and constitute the definitive Revelation to all men. The Ten Commandments are complemented by the Eigth Beatitudes and by the Evangelical Counsels. Peter, (the head of the apostles, appointed by the Savior himself), and Paul (the apostle of the gentiles) die in Rome under Emperor Nero in the year 64

A.D. Persecution against the Christians broke out in the 1st century and accompanied the Christian Church since then. As Saint Augustine said in the 4th century, Christianity spread out with miracles or without miracles. But, if without miracles, that was the greatest miracle. From very early times, spiritual life in Christians was nourished by the Sacraments (Baptism, Confirmation, Holy Communion, Confession, Matrimony, Holy Orders and Unction of the sick). Local churches were presided by their Bishops, as successors of the twelve original Apostles, and were supposed to remain in communion with Rome's Bishop, successor of Peter.

Today here are three broad divisions within Christianity: Roman Catholic, Orthodox, and Protestant. The Orthodox church was born from a schism in the 11th century, but essentially it shares in common the articles of the faith with the Roman Catholic Church, except recognizing the primacy of the Pope as successor of Peter. The Protestant churches, which are many and varied, separated from Rome since the 16th century. In general they reject apostolic tradition as a source of Revelation and declare "sola scriptura" as the only source. A strong movement has developed in the 20th century, sponsored by Pope John XXIII and by the Second Ecumenical Council, seeking the reunification of Catholics, Orthodox and Protestants.

Islam. According to the Encyclopedia Britannica ((Book of the Year 2013) here are 1.609 million muslims in the world today (22.8%). They are growing in numbers faster than any other of the great world religions. Islam means "submission" in Arabic, and refers to the attitude required from believers before God. Muhammad (570-633 AD) was the prophet who preached successfully " submission" to God (Allah) in his native land, Arabia. From there, Islam spread quickly through West Asia, North Africa, and, within only eight years, it arrived to Spain, in Southern West Europe. In the 15th century the Ottoman Turks managed to put a foothold in the Balkans. Today, the only great continent in which they have

REASON, SCIENCE & REVELATION

only very few adherents is America.

Muslims are respectful of the Bible. They accept the Ten Commandments and consider Christ a great profet. According to Islam, God reveled himself finally to Muhammad in the Koran. This sacred book does not contain a systematic doctrinal exposition, but a body of beliefs, practical duties, rituals and laws that have emerged from it. The believers constantly praise and glorify Allah. He is awesome, transcendent, almighty, just, loving and merciful. He alone deserved to be prayed to and adored. Muslims seldom ask favors from God, limiting their prayers to thanksgiving and adoration. Associated with God are angels ("messengers") and prophets. According to Islam, Muhammed was the last great prophet. The five principal prophets were Adam, Noah, Abraham, Moses and Jesus. Jesus, born of a virgin, performed great miracles. Muslims, however do no accept the crucifixion and they believe that Jesus was taken by God, leaving only a shadow in his place, a view which was common amongst the Gnostics in the early Christian centuries.

.

9. Creation
by Manuel M. Carreira, SJ.

The desire to know *"where we are, where we come from, and where are we headed"* is a necessary part of our endless search for Truth. In that series of questions one can also see a logical and time-like chain of reasoning, that from the knowledge of our present state argues to an explanation found in previous conditions, to an always more remote past where causes have to be sought for succeeding effects. But this process cannot be pushed logically to an infinite series of steps: we must finally inquire about an origin *before which there was no "before"*, even if this statement seems to imply the paradox that time itself must have had a beginning.

Mythologies and philosophical ideas of those cultures that are historically known, avoid the endless series -seeking always previous causes and states- by postulating a basic element, eternal and self-sufficient, of a material nature. The Earth –in an all-encompassing sense that includes also the visible astronomical bodies- is considered to be the root from which everything sprouts, even the gods themselves, who are frequently just a divinized reference to the basic elements: the oceans, the heavens, the winds, fire or water. Those gods can have their own rivalries and fights that are in some cases used to explain the different levels of structures that end up with the making of human beings.

Even if the final state might refer to the gods as immortal (for instance, in the Greek Olympus realm), those divinities are still subject to ignorance, pain, defeats and subjection to the more-or-less arbitrary power of other gods. They are shown as "superhuman" heroes, rather than divine entities in a transcendent sense, and they show vices and behavior that would be unacceptable in human society. Their power is also limited, in the basic sense that they need some kind of raw material to make something useful for their purposes; there is no mention of "creation" in the philosophical sense of going "from nothing to something", a step that requires an unlimited Power. Even the continuing existence of the gods obliges them to seek food, either through the offerings of sacrificial victims by the faithful or by seeking some type of different nourishment in their abode.

It is in this cultural setting where the Jewish people provide a unique and new way of understanding the divinity, that is not identified with anything visible, and that is free from any limit of space or time, unchanging, all-powerful, endowed with infinite wisdom, holiness, justice and generosity. This portrait is given a concrete expression in a poetic story, that has many elements in common with cosmological views of other peoples of the same area, but that is free of any materialistic conditionings suggested by the human experience. It is in the Genesis account where we are taught –in an implicit parable- that God is a loving Father, that He is the only sufficient reason why everything else exists, and that all creation is the work of his Wisdom and Love.

This is a radically new teaching, presented in images and words that are suited for a pre-scientific society, but that retains its value in our space era. It isn't meant to give answers to questions belonging to Astronomy or Biology, something that has been wrongly sought in previous times trying to obtain a "concordance" between the Bible and modern science. But it stresses the ideas of *order, harmony and intelligibility* in the entire realm of existence: those properties without which the very science -that we are so justly proud of- would be impossible. This same science enriches our understanding of the Genesis account, furnishing a variety of details that lead to a greater appreciation of the work of the Creator.

In a letter of Pope John Paul II to the Director of the Vatican Observatory (dated June 1st, 1988) we can read: *"If the cosmologies of the ancient Near Eastern world could be purified and assimilated into the first chapters of Genesis, might contemporary cosmology have something to offer to our reflections upon creation?."* In my opinion, the answer is very positive: it is the most up-to-date cosmology the science that best underlines de marvelous work of the Creator.

The initial setting of the genesis account

The first words of the book of Genesis, *"In the beginning God created the heavens and the Earth."* include –in the standard translation- the idea of *creation* that we understand with the insights of twenty centuries of Theology. But it is possible that a more exact version would be: *"When, at the beginning, God made the heavens and the Earth…"* even if the Hebrew word *barah* appears only for God's creating activity. The first clear reference to a strict creation from nothing seems to be found in the second book of Maccabees, but not in Genesis. It was not an important concern of the primitive Hebrew society whether the world was or not eternal; rather, its unique and total dependence from God had to be clarified by showing how everything was fit for our existence due to God's providential care. This is the reason why the Genesis account implicitly stresses the idea of *God as the source of order*, of a wise series of steps meant to allow human beings to exist and to reach their full development.

The implicit parable describes God as a Father who plans a home for his children, and who proceeds in a systematic way, from the general structure of the dwelling to the details of comfort and beauty that make a house into a home. Man is presented as a living Image of the creating Father, his representative and helper, meant to preside upon and develop the work that is given with a totally unselfish Love. Nothing is asked as payment for the gift, except the logical attitude of thankful reverence for such generosity, as we find expressed - millennia later- in the *Contemplation to Attain Love* that closes the "Spiritual Exercises" of St. Ignatius.

It is worth noting that the God of the Bible has no special personal name except the simple expression of his divinity. He is the only true God, and thus it isn't necessary to identify him with a proper name as it would be expected if there were several deities. No reason is offered to explain his origin: He is already present at the true beginning of everything, not only in a previous time, but more deeply as the ultimate reason in the orders of causality and sufficient reason.

The initial state of anything that is not God is described as a "chaos", where disorder, darkness and emptiness sum up the total absence of any positive quality independently of God's action. Water itself –the basic requirement for organic life- is presented as an unlimited abyss that Man cannot encompass, dreadful in its blackness and turmoil. It is a way of thinking that appears deeply ingrained in the subconscious mind of those nomads of the desert, always afraid to leave the solid ground to venture into the seas, even when settled in their cities and farms, as we read frequently in the Psalms.

But upon that limitless ocean there hovers, as a Power that can bring it into subjection, the Spirit of God. A Spirit that is also a living breath and that expresses the powerful will and the wisdom of the biblical God, *a living God*, not dead like the idols of the pagan cults of other peoples.

Laying the basic structure: the first three days

The First Day: Light and daily rhythm

The first logical step in the working plan to develop a habitable structure will be to provide light: nobody works in the dark. The physical nature of light was totally unknown (until very recently!) and there is no attempt to specify a process that will obtain light from some kind of fuel, or to link its existence to astronomical sources. Light is an independent and previous reality that comes into being through a simple command of God: "*Let there be light!*", and there was light. As an artist satisfied with his work, "God saw that the light was good, and he separated it from darkness". Thus begins the orderly succession of day and night, a basic framework for human activity in the experience of those peoples for whom the Bible was written. *It is the First Day.*

Our 20th century science has found, with new technologies and new ideas –not immediately recognized as valuable- the first "light", previous to the Sun and the stars, in a "beginning" that can be described as chaotic. An unimaginable cauldron of dense and powerful energies was the birthplace of particles and atoms of hydrogen and helium, synthesized during the first 20 minutes of the Universe. Against the widespread prejudice that the Universe had to be eternal, it is the weight of scientific data that forces us to accept a beginning with no "before" that can be described scientifically.

That initial seed of super-dense matter, with parameters that define its properties with extreme precision (reaching 50 decimal places in some cases) undergoes an expansion during the "first cosmic day" when the light of the Big Bang is followed by the night of a sky without stars during millions of years. We can still detect and analyze the minimal "heat" (at 3 K, almost absolute zero!) of that great fire that marks the start of cosmic history 13,700 millions of years ago; we have also found its "ashes" in the correct abundance of hydrogen, deuterium and helium (calculated by Gamow in 1948). This is truly a first "day and night" in an unimaginable level... because God is not subject to time and he has no waiting periods during the creation process.

The key discovery of modern cosmology was based upon Einstein's equations (that Lemaître fully developed) and the work of Edwin Hubble at the Mount Wilson Observatory. The *General Theory of Relativity* required the acceptance of a "finite but unlimited" universe, expanding or contracting, instead of the static, unchanging and eternal model uncritically accepted by the majority of astronomers a century ago.

We are entitled to say "uncritically" because the obvious objections were not faced: an infinite universe, with an infinite amount of mass in all directions, would cause the gravitational potential to be infinitely high at each and every point in it, thus ruling out the potential differences that give rise to net gravitational forces. And an eternal universe would have now only dark stellar corpses everywhere, since in an infinite time all fuel sources for nuclear reactions have to be exhausted. Only a strictly understood "continuous creation" of new matter could solve the objection: a solution proposed in the 1950s by Hoyle, Bondi and Gold, but incompatible with well established data.

The discovery of the background radiation in 1965 by Penzias and Wilson (who shared a Nobel prize for their work) and the abundance of quasars only in the very old universe, firmly established the logical consequence of the expansion announced in 1929 by Hubble: the known universe began in a state of extremely high density and temperature. Its initial conditions –including the parameters of elementary particles and forces- determined the evolution from the primitive "chaos" to the majestic structures that we now study with the instruments of the space era.

Second Day: an environment free of water

In the biblical account we find now a description of the second day that can make us smile at a cultural background that seems to be proper of a child's tale. Up to this point we were told only of an enormous watery mass, unruly and lifeless, as if it were an enormous blob without container or barriers of any kind. It seems necessary to open up a hollow space in it, to have room –like a cavern- for air and hard ground where something can be built.

The obvious experience of rain that comes from above, and of underground water flowing from wells and springs, suggested some arrangement of "waters above and waters below" that seemed to require some kind of dome, separating both locations. This is the work done on the second day: the "firmament" is created to hold the waters of the upper realm, keeping them from falling down (naturally!) and flooding the low areas. This "glassy" dome, invisible but strong, is found also in the cosmologies of Egypt and other nearby cultures; its role is previous to the presence of astronomical bodies, and if one suffers the "Flood" it is because the trapdoors of that dome were open. It isn't clear to what an extent this description was considered to represent a real thing or just a poetic fiction: it could easily be seen that rain would fall from low clouds, not from the stars. But the image was commonly used as we have it here.

Nothing like that can be found in any scientific description, either of the universe or of our home planet. We should rather say that the primitive Earth most likely looked like the Moon: a barren and dry rock covered with craters. We attribute the oceans to the ices of millions of comets (condensed in the pre-planetary nebula 4,500 million years ago) that impacted the early Earth and most likely gave it an unbroken layer of water that lasted millions of years. Those "waters from above", and the volcanic venting of underground gases that included water vapor, gave the possibility of harboring life to our planet, where water has been present in all three states –solid, liquid and gas- during billions of years, in an environment that is unique within the Solar System.

Water is possibly the most surprising and fertile substance, due to its physical and chemical properties. It is an almost universal solvent –there is gold in the ocean waters- and it is less dense in the solid state than as a liquid when frozen under normal pressure. Thanks to this unexpected behavior, ice floats, and the oceans are not a solid block from the bottom up, making life impossible. And nothing is found comparable to water in its suitability for the full activity of the carbon chemistry (*Organic Chemistry*) that is the only possible basis for the enormous complexity needed for the DNA molecule, and for the metabolic paths required to build a living structure.

We can logically retain the idea that stresses that the cosmic abundance of water in the heavens (in the solar nebula) and in the interior of the Earth's crust is crucial to make our home suitable for life. We do not regard water as a menace, as long as there are limits to its presence and its activity (even in the desert, an oasis is pleasant because it offers needed water.) The biblical account is not old fashioned in that respect as long as we see it with our present understanding of astronomy and biology. This would be the meaning of our scientific *second day* that receives again the approving remark of the Creator: "*And God saw that the firmament was good.*"

Third Day: A fertile ground and vegetable life

On the third day we are told about a firm and solid dry ground, the place where a human habitation can be set up. God limits the expanse of the waters, confining them to the basins of the oceans, and this frees the dry land of the continents. Without any knowledge of geology, the imposition of borders to the waters is seen as the effect of a divine command that restricts the presence of the oceans by assigning to them a proper place they must keep, not by any physical reason. Those confining borders are considered as everlasting (they seem to be immutable from the viewpoint of human experience) and thus they insure a habitable space, where mankind can develop without the fear that the oceans —always dreaded and prone to chaos- might invade the area assigned to Man, with catastrophic results. This happens when there is a tsunami, as we now know, but the biblical writer was not aware of such a phenomenon.

Modern geology has confirmed the daring hypothesis of Wegener, explaining the origin of changing ocean basins through eons of evolution. Continental plates –pieces of the crust- are pushed by convective flows of slightly fluid rock of the mantle, heated by the reservoir of the metallic core (iron and nickel at over 4000 degrees, comprising one third of the total mass of the Earth.) The pressure of those flows breaks the rigid and thin crust (only about 30 km. deep) and causes the plates to collide to form mountain ranges, and to give rise to volcanic and seismic activity in areas where one plate overlaps another (a process known as *subduction*.)

The continental plates were fused together 220 million years ago in an enormous continent –Pangea- that later broke into two (Laurasia and Gondwana) with further evolution raising the Himalayas when India collided with the southern flank of Asia. The Atlantic basin opened up with a long crack running from Iceland to Antartica, and we can still check the growth of this basin –at about 3 cm. per year- with the help of laser beams, sent to reflectors placed on the Moon by the Apollo astronauts.

This slow and persistent process is vital for the renewal of our land and atmosphere, whose gases react with the minerals of volcanic sources. The weathering of rocks, and erosion processes, recycle surface materials and clean the atmosphere of excess carbon dioxide, limiting the proportion of oxygen in the air as well. Nothing comparable has taken place in the other planets, similar in structure to the Earth, in our Solar System. Our planet is the jewel of the Sun's family, the only one suited for life.

Once the basic structure is ready, where Man's abode can be set up, God begins the more detailed process of making sure that it has all the necessary things for human life. The first and most obvious need is that for food, and this can only be found in lower levels of life. Not having microscopes, it seemed that the most elemental form of life is a plant, devoid of sense impressions and of the mobility and freedom of any animal. It is logical, therefore, that God will first create plants upon the earth, that is now free of the oppressive presence of an all-encompassing ocean. And so it happens, but with a new way of obtaining the desired effect: God orders its creation to contribute to the development of his plan with its own activity: *"Let the earth produce* green grass, seed-bearing grass, and fruit-bearing trees, each one according to its kind".

Once more, the language is used in a way that stresses order and hierarchical levels in creation. The green grass – without flowers or visible seeds- will be food for the cattle and other animals especially useful for man. "Seed-bearing grass" refers to cereals, the source of the basic food, bread of different kinds. And the "fruit-bearing trees" -that also have a seed that ensures their continuing supply- offer the multiple richness of grapes and wine, olives and oil, dates and figs that were a staple of the nomad's fare and of the settled Israelites. Everything appears as the effect of that command that allows inanimate matter to organize itself in the marvelous biological system that constitutes a plant, capable of using minerals, water, carbon dioxide and sunlight to synthesize carbohydrates, sugars, lipids and aminoacids. Such a variety of chemical processes is still an unending source of wonder and new discoveries in our own time.

The biblical writers did not know that green plants are the key element in explaining how the primitive atmosphere of our planet became rich in oxygen, or that without their continuing support very soon the Earth would become uninhabitable. In sedimentary layers of 2 billion years ago we find evidence of extensive mats of microscopic green algae, the reason why 600 million years ago there was a sudden "explosion" of living forms, when the oxygen level in the atmosphere -similar to the present one- allowed the development of macroscopic life.

Two evolutionary steps, both impossible to foresee and of almost zero probability even in cosmic time periods, changed the metabolism of single cells that had developed without oxygen, and for which this gas was a poison. The first one enabled some cells to synthesize organic components by the use of chlorophyll and sunlight. When the abundance of oxygen released in that reaction reached a sufficient level, a second genetic change allowed some cells to utilize oxygen for a more efficient metabolism, and thus prepared the way for multi-cellular life, both vegetal and animal.

It is the difficulty of making even an educated guess for those changes to occur by chance anywhere else that has led serious scientists to consider that microscopic life might now exist in many other places of the universe, but that something as complex as a mouse -not to speak of human life- is almost impossible to expect outside of our planet. Even here, the development of life seems totally unpredictable and not to be expected to follow the observed path no matter how many times one could start it.

With the green carpet of vegetal life covering the continents, the third day comes to its completion, and God again gives his approval to the state of things as the third night begins.

Preparing things in detail for human life: the next three days

A second period of three days parallels the first stage of creation. To the first day, marked by light, corresponds the fourth, when heavenly bodies appear. If the waters mark the second day, life in the oceans and the air flourishes on the fifth. And the dry land covered with vegetation (3rd day) has prepared the environment for animal life on the continents. We can study those developments more fully.

The Fourth Day: astronomical lights

Three days have so far been described -with periods of light and darkness- *without any reference to astronomical bodies* to determine those changes. It has been said, almost as a joke, that for the primitive Israelites the Moon was more useful than the Sun, because it came out at night, when it was needed, while the Sun was there when it was already light. The role of the heavenly bodies was mainly to *preside* with their impressive presence over all the affairs taking place on earth: they were considered to be lamps, moving or fixed to the heavenly dome, just as the stars that form the various constellations. They did have an important role –especially from the viewpoint of a religious calendar- by clearly determining day and night, seasons and yearly cycles.

This is the "host of the heavens" -the visible servants of the Creator and of Man- providing an unchanging basis for order and times, but without being divine in any form. Thus the biblical account rejects any worship of the astronomical bodies, something that was openly present in Egypt and, in a more or less obvious way, in many other cultures of the Near East and of other peoples throughout the world. We find later in the Psalms the insistent teaching about the nature of *servants* of the Sun and the Moon, chastising those who from the rooftops are observing the heavens for changes in the motion of the planets. This "astrology" –a caricature of scientific astronomy- still appears today in the childish horoscopes of our mass media, in spite of the well proven lack of any foundation or predictive ability. The biblical expression referring to God as "The Lord of Hosts" is not a war-like banner, but a reference to the "host of Heaven" of luminaries, great and small, that move following God's will in an ensemble that is orderly and lasting.

Modern science has answered the question of *where we are*, by giving distances and orbits within the Solar System, and next by opening immense vistas, first when distances to stars could be measured (by Bessel in 1838, using the parallax method) and then by determining that the Sun with its family of planets is only a "citizen" in the enormous "cosmic city" that we call the *Milky Way*. Its census would number over 100 billion suns arranged in a disk that light crosses in 100,000 years of travel. Our orbit around the center (which is only observable with infrared and radio waves) needs 250 million years to complete a turn. No wonder that a century ago astronomers felt overwhelmed by such immensity and thought that they had found the true extent of the whole universe.

But new telescopes of varied technologies soon established that -at distances that make the size of the Milky Way, our *galaxy*, insignificant - one can find and photograph other galaxies until the sky is so incredibly thick with them, even if they appear as simple specks of light, that the entire universe might contain as many galaxies as there are suns in the one we inhabit. Nobody could have dreamed at the beginning of the 20th century the distances of billions of light-years and the overwhelming abundance of suns and planets, so far away that even the largest telescopes can show the closest stars only as dimensionless points of light.

Still more amazing is the knowledge acquired to explain the evolution of matter from the explosive beginning of the Big Bang to our age. The Earth and our own body are, literally, made from stellar ashes added to the primeval hydrogen. Only through the synthesis of carbon, oxygen, calcium, iron, silicon... in previous generations of stars can we account for the existence of our planet and of the ingredients that make life possible. The humble servants of the divine plan in the "host of Heaven" prepared the "clay" of our bodies, and even that of the Body of God-made-Man in the Incarnation.

A great English astronomer, Eddington, commented around 1920 that we should be able to understand *something as simple as a star*. The theory of stellar evolution is, in fact, the most satisfactory and complete section of present-day astronomy, even if there are many details that need to be clarified, especially when we deal with the final stages of evolution of very massive stars. We can describe in simple terms how a sufficiently massive cloud of gas –dense and rather cold- should contract under its own gravity, thus compressing and heating the gases in its core. When the temperature of 10 million degrees is reached, nuclear reactions take place (as shown in our laboratories) that synthesize helium from hydrogen and release energy that counters the weight of the outer layers. As long as there is a sufficient amount of fuel to maintain this balance, at increasingly high temperatures, the star avoids gravitational collapse, but gravity always wins at the end.

Sun-like stars die a relatively quiet death, slowly losing their outer layers until the hot corpse at the center (made of crystallized carbon in a sphere the size of the Earth) is bared. But stars of considerably greater mass evolve -rather rapidly by astronomical standards- until their internal instability causes an almost instantaneous collapse of the main share of the stars' mass, compacted into a sphere just a few kilometers across. An explosive bounce throws the rest into space, with all the materials previously synthesized, and during a very short time (perhaps hours or minutes) allows the production of the heaviest elements. This was the source of the metals present in the cloud that 4.5 billion years ago gave birth to the Solar System.

The cosmic ages, with their unimaginable long periods of evolution, were not a waste of time, and the overabundance of so many heavenly bodies that Man will likely never visit is not useless. Once more we should enthusiastically echo the approval of the Creator at the end of the fourth day: *And God saw that the heavens -with their variety of "stars"- and the fertile ground also, with its living carpet of grass and trees, were good.*

Fifth Day: Animal life in environments not controlled by Man

There is now a more rapid and varied rhythm of activity, a burst of creative power that makes life develop in multiple forms in the two environments —water and air- that are both most prone to change and less subject to the activity of man to control their resources. God uses once more the powers given to nature, ordering the waters to "teem with animals" as varied as oysters and whales, jellyfish and giant squids, humble fish of lakes and rivers, or unknown monsters supposed to move in the dark depths of the oceans.

As a second panoply of agility, beauty and fantastic shapes, the world of birds is also brought into being, from the small sparrows -almost invisible with their dusty color and insignificant size- all the way to the majestic eagles that look at us from their lofty heights. There is an all-encompassing presence of life that will be permanent, due to the divine injunction *to be fecund*, something that is presented as the first *blessing*, not a simple approval from the Creator.

The Bible especially underlines the power to give life when it speaks of the *Living God* as opposed to the dead and lifeless idols of the pagans. There is no hint in the books of the Old Testament of the amazing mystery of the Trinity, that implies the total giving of the full divine nature as the essential activity of the divine Persons, but the living nature of God is always mentioned, even if life without time intervals is incomprehensible to us. God is described without any reference to a possible origin of his existence, unchanging through the ages ("for You a thousand years are like a day") and can only be said to have all possible perfection in his infinity. But because God is mysteriously living, he can communicate life, thus showing his true nature in a more direct way than by creating the astronomical world, no matter how marvelous it actually is.

Elementary family relationships, already present in the way birds care for their chicks in their nests, point to deeper and more significant ties to be expected when the highest degree of life appears on Earth.

Biological sciences study the development of life, beginning with an unknown start in the waters of 3.5 billion years ago. We do not know how or when or where the first cell actually appeared: the complexity of any cell is such that to try to explain it in terms of "chance" is tantamount to saying "just because". Chance is not a physical force, cannot be detected in an experiment, and it can never be the reason for order and complex structures.

All we can say is that life did appear, possibly in a small puddle by the shore of an ancient shallow sea, or in some hot spring or volcanic vent in the deep ocean. Crystals or clays might have acted as templates to align aminoacids that lightning would produce in a primitive atmosphere, leading to a self-reproducing macromolecule. In any case, that first life was not vegetal —we define plants by their ability to synthesize new organic materials- and it depended for food upon the presence of other simpler organic compounds in the waters surrounding it. We talk about facts that seem to be logically unavoidable, but we do not have a satisfactory explanation for them.

It has been 50 years since Urey and Miller, in Chicago, mixed gases likely to be present in the first atmosphere of the Earth (water vapor, methane, hydrogen, carbon dioxide and monoxide) and subjected them to electrical discharges over a period of hours, obtaining some of the basic "building blocks" of a living organism. More refined ideas about the primitive "air" have led to different experimental results, with a greater variety of useful compounds, possibly formed in a volcanic environment. But nothing has happened since to advance the process to anything approaching the simplest cell. And while St. Thomas would accept what we call "spontaneous generation" as the expected flowering of potentialities, given by God, to have matter develop life when set in the proper circumstances, it is still true as a scientific fact what repeated experiments (of Pasteur and others) stated as an unbroken rule. *"Omne vivum ex vivo"*, all life comes from previous life. We are as far from the synthesis of a cell as were the philosophers of the Middle Ages. In a more or less open admission of this fact, it has been proposed that life should have come to Earth from outer space...where it developed in unknown surroundings.

For the first 3 billion years all life on Earth was microscopic, with single cells that could not tolerate oxygen (anaerobic.) Only green algae –also single cells- that covered large areas in the oceans- could provide the oxygen abundance that allowed the "Cambrian Explosion": a brief period in the geologic calendar (only a few million years) when all the basic types of present life forms suddenly show up as fossils in sedimentary rocks. It is still a mystery why such a variety -of phyla, families, genres and species- blossomed first in the oceans, and then in coastal lands: fishes, amphibians, reptiles, birds. And this evolution was conditioned by five great extinction events, due to astronomical or geological changes, so that only about 10% of all living forms have survived.

The basic unity of all life forms on Earth is scientifically well established: the same aminoacids, the same symmetry of organic molecules, the same way to encode and transmit genetic information, are reason enough to accept it. But it is just as true that it is extremely difficult to explain in detail the evolutionary process, and that Biology cannot give a satisfying reason why specific changes took place, even if they are attributed to genetic mutations that appear as random events.

We should remember that the scientific method cannot *experimentally* prove or disprove that there is finality, even when dealing with obvious products of human technology. The question about choosing between *chance and intelligent design* belongs in the philosophical realm, and must be answered by metaphysical arguments that do not belong in the biological description of evolution. In biology we must ask *how* evolution occurs, but not *why* did it occur the way it did.

The fossil evidence allows us to recognize fishes as the most primitive vertebrates, followed by amphibians and reptiles, whose development reached its peak with the dinosaurs and their overwhelming presence during 150 million years. An ecological disaster –almost certainly triggered by the impact of a 10 km. rock (an asteroid) upon the Yucatan peninsula- put an end to their reign 65 million years ago. Their humble descendants are the present-day reptiles and the birds, and it is in this less threatening environment where evolution could develop further from the first mammals.

Sixth Day: Land animals and human life

For those who were instructed by the books of the Bible, people originally nomadic even if later settled in farms and towns, wealth meant domestic animals: sheep and goats, cattle, donkeys and camels. These are the animals that are first meant when the next step of creation is described on the sixth day, but reptiles are also included: they were encountered mostly in the desert as snakes or lizards, and crocodiles were known to exist in Egypt and in rivers of the region,. This off-hand mention of reptiles might be related to the idea of their *impure* nature that makes them unsuitable as food, and that could be due to pagan cults -where serpents were symbols of fertility goddesses- or to idols of Egypt, tied to wild beasts, perhaps absent in Palestine.

Without further detail, "all other land animals" are mentioned as well, to make it clear that all living things exist due to God's all-powerful word.

Paleontology gives only a partial description of the evolution of mammals, mostly after the dinosaurs disappeared. Our stock of fossils is always incomplete, but we have enough to follow the evolution of the ancestors of the horse through animals closer and closer to the present beast. This also applies to primates who are biologically closest to Man. But it is difficult to find convincing intermediate steps to go from the hypothesized forebear of the whale –a land mammal the size of a donkey- to the imposing streamlined monster we observe today, the largest animal in the entire history of earthly life. A mammal still, it can dive deeper than a nuclear submarine and can stay underwater without breathing for over an hour, has a heart that pumps 1000 liters of blood in each contraction, and can store oxygen in a thick layer of fat under the skin. We really are left with a mystery when we want to explain such evolution.

It is also difficult to explain the instinctive *program* that determines the way a spider weaves its web, or how a bee makes the honeycomb, or a bird builds its nest and seeks its food. How far can an instinct –a program that is genetically inherited- lead to mimicking behavior and learning, as found in present-day primates? In the animal world there is no culture that can be shared with a system of symbols, whether visible or expressed with sounds, nor do we find any hint of a conscious self-knowledge or a free decision. We can refer to animals as "biological robots", astoundingly complex but with the limitations inherent to the lack of true choice among alternative ways of acting that can follow a free decision. For this reason we cannot talk about animals in terms of personal responsibility, with rights and duties.

At this point the Bible presents the Creator looking at his work and saying again -but for the last time- that he approves of what he has created and that cannot be in any way different from what he has decreed: "And God saw that it was good", this world brimming with vegetal and animal life.

The creation of Man

Once the setting is completed, the provident Father –who made the proper home for his children- caps his work with a new creative act carried out with special solemnity. Not by pronouncing again a distant "Let there be…" or by ordering the waters or the solid ground to use their innate powers to make a new life form sprout from them. God will play a unique personal role in the creation of Man, who will be his "Image and Likeness", a living being sharing in the activities proper to the intelligent and free Creator. Therefore Man will be a "son" in a very special way, since a son is the living image of the father.

The word "Elohim" that is used for the divinity in the original language of Genesis is a mysterious plural form, and it seems unsuited for a text where one sees how the exclusive uniqueness of God is constantly stressed. But it is also a plural form that is used here to announce the new creative work, just as if a common plan were decided by several persons of equal rank: "*Let us make Man according to our own image and likeness.*" And his special dignity is then clearly stated: he will be God's representative with authority over the created world: "*To have dominion*" over all living things created previously, as a son and heir who is never listed among the assets that constitute the material wealth of the head of a household.

This new order of existence, in a level between the Creator and his previous works, is stressed with a meaningful insistence, almost showing the amazement of the writer: "*And God created Man as his image, he created him in the image of God, and he created them male and female*". The difference in sexual identity does not imply a different nature or dignity, since the human couple appears from the first moment as coming out from the hand of God with that likeness to God that gives each human being a unique place in the whole of creation. There is no discrimination based on race, caste or sex in this view, so positive in its simplicity, of our common nature.

When creating the first human beings we do not find the usual approval that states the goodness of the work: an omission that suggests that the goodness of Man will be dependent upon our *freedom*, not fixed by our biological structures. There is, instead, the blessing of cooperating with God to be sources of new life, a blessing given before to land animals, but that in this case is joined by a new declaration of Man's dominion: "*Be fruitful and fill the earth, subject it, and have dominion over the fish in the sea, over the birds in the sky, over the cattle and over everything that lives and moves over the earth*". Everything has been made for Man, and Man only for his Creator: there is nothing in the created order that might have a right to deprive a single human being of a dignity that is always above any concern about usefulness for another, be it an individual or a society.

Our science cannot but recognize that there is a qualitative difference –not only in degree- between Man and the rest of living things on our planet. Every materialistic effort to explain what we are, in terms of genetic codes or brain complexity, fails when faced with undeniable facts that are incompatible with those apparent explanations. The size of the brain cannot be directly related to intelligence, and identical twins with exactly the same DNA do not behave the same way, even when linked by common organs that necessarily oblige them to be exactly under the same environment. Each one is a different person, with diverging tastes and desires.

At a deeper level of reasoning, we have to stress that

abstract thought –the reason why science, mathematics, poetry, are possible- cannot be due to any of the four forces that are used for a physical definition of matter. It does not produce anything with experimentally measurable properties, and no scientist admits that it will have a physical effect upon the material surroundings. Nothing is gained by claiming that we already have "artificial intelligence": even if we were to accept as possible that a set of the electronic circuits of a computer could arise by chance through chemical evolution, we would still need the *software*, the "program" that can give purpose and meaning to the electric currents going through the circuits. Such program cannot be the result of chemical reactions and it would have no meaning –no informational content- unless there is an arbitrary system of symbols expressed by the flow of electrons through chosen paths.

Our likeness to the Creator, so eloquently proclaimed in Genesis, cannot be due to our bodily structure, but only to *our rational power and our free will*. This cannot be explained by genetic evolution, even if it is acceptable to admit that - through evolution- God prepared matter to make it suitable to receive a *spirit*, the only sufficient reason for that kind of activity. As long as science remains within the bounds of its own methodology it cannot argue against this view, even if the interaction between mind and brain remains a mystery. Nothing has science discovered to solve the problem, in spite of the efforts of many centuries, except the recognition that some "mental" disorders can be due to chemical imbalances or to brain tumors.

To use a sobering comparison from physics, we can also confess that it is a mystery how elementary units of matter can be observed both as particles and as waves, and how the two mainstays of 20th century physics, the *General Theory of Relativity* and *Quantum Mechanics*, can be combined in modern cosmology. We have numerous and undeniable data that make it necessary to accept those different ways of acting for matter, but we do not understand how things can be so.

This first chapter of Genesis ends with a poetic dream: all

REASON, SCIENCE & REVELATION

animal life will be strictly vegetarian. Something hardly applicable to fish or to animals that cannot possibly digest vegetable matter, as is the case for a lion or even a humble mosquito or a leach.

It is wrong to think that the need for meat is due to human sinfulness, as if the suffering of animals that are the victims of other animals were our fault. Millions of years before Man, we find the most terrible carnivorous predators among the great dinosaurs. It is a common mistake to speak about animals in human terms, assigning to them rights and duties and thus classifying them as ethically "good" or "bad".

Less acceptable still is the view that an animal has a greater dignity than Man, leading either to adore a beast or to place its wellbeing ahead of a human need. This is the implicit thought behind oriental tenets regarding reincarnation, giving human status to any animal in our surroundings.

The full description of the creative process ends with a remark that sets a logical basis for the observance of the Sabbath as a religious day of rest: the seventh day is especially blessed and made holy by the Creator taking a rest at the completion of his work. Not because his activity imposes a need to regain strength –a simple "Let there be…" is all that is needed to bring into being anything he wants- but because Man will need a rest at regular intervals to avoid the crush of his own work. No tiresome activity should make us forget our relationship with the Creator at a personal level.

The seven days of the week had their origin -from a mythological mixture of astrology and poetry- in the seven "travelers": Sun Moon, Mercury, Venus, Mars, Jupiter and Saturn. They are the heavenly bodies that, with the naked eye, can be seen to move against the background (the "fixed stars" that make up the constellations.) The Genesis view gives those days a deeper meaning: there is a previous order – before any stars or planets- in the providential action of God preparing the world for Man.

This is why Christ said, much later, that *"The Sabbath was made for Man, not Man for the Sabbath"*, a remark that is applicable to all external norms that might end up by limiting even the possibility of doing good to others.

Thus the first chapter ends its poetic description of a *beginning*, that is couched in language that even the most primitive society could understand, and that set forth the main ideas that distinguished biblical theology and cosmology from any other human effort to answer the deepest questions we all face. The language of a primitive world view did express important ideas that our present science makes even more cogent. They underline the Wisdom and Goodness of the Omnipotent Creator and, as a consequence, the dignity of His masterpiece: the Thinking Animal, Image and likeness of the Creator.

Second account of creation: Man as center

The second chapter of Genesis gives a different account of creation, more synthetic and without a specific time schedule or distinction of stages, because it is centered upon human activity and our response to the Creator who made everything for our benefit.

In this more theological view God gives Man a body made directly from clay, working it with his own hands, like a sculptor that makes a human image with sublime skill. He then infuses into this sculpture a "breath of life", a superior reality different from the inert clay, and it comes to life. But its surroundings are barren and totally devoid of the essential means to sustain life of any kind, *"because Yahve had not sent as yet rain upon the earth and there was no Man to work it"*.

But now that Man is present God makes a paradise where he can happily live, "with trees beautiful to see and offering fruits that delighted his palate", thus joining beauty and usefulness. The most important of those trees were two that have no counterparts in any modern botany: one, the "Tree of Life", was the guarantee of immortality; the second, "The Tree of Knowledge of Good and Bad" was meant as a symbol and test of the fidelity and obedience due to the Creator.

It would be a total misunderstanding to consider the

prohibition of eating the fruit of the Tree of Knowledge as a limitation of the desire to know, that is a necessary consequence of our rational nature, always seeking Truth, Beauty and Goodness. In the Semitic language "to know" frequently includes the meaning of "to control", to impose some kind of dominion, even when the male "knows" the female joining her in starting a new life. In pagan environments of that time, and even in magic rites of our own age, *to know the name* of some superhuman agent is the key to control it and to extract favors. It is this "knowing" that Adam should not seek, under pain of death, since the attempt to control God would constitute a rebellion, acting from a position of power, not with the subjection proper of a creature. "To *know* Good and Bad" means also to be *an independent arbiter* of what is allowed and what is not, defying God..

Two other ideas are especially significant in this second version of our origins, where Man's role is more explicitly developed:

First, it is stated that man should not exist in an isolated solitude: being a *person,* he *needs to relate to other persons.* God makes all animals of the ground and of the sky go past Adam in a parade of incredible variety, and he exercises his dominion upon them assigning a suitable name to each. Once more, *knowing* them and determining their way of acting, since the name –in this frame of mind- foretells the future role of the one being named: we can refer to Christ giving a new name to Simon Peter.

But among all those animals none is found *in the likeness of man*, even if it might look similar in a purely physical way, and thus none can be *an adequate companion* for Adam. We were insistently told earlier that Man is made in the *likeness of God* – even if God is not described in any way that might imply a bodily shape- and now the likeness is denied with respect to living beings that do not seem to be that different in looks. It is a simple -but profound- way to stress once more that human dignity is unique and based upon our rational nature.

Second, the woman -Eve- is not independently formed from inanimate clay, but rather from the close intimacy of Adam's heart, as his "flesh and bones." This original unity is the guarantee of the singleness of the human species, and it also establishes the equal rank of the woman as the *adequate companion* to the male, not as a pet or domestic animal. And their mutual attraction will be such that it will prevail upon family ties with the parents of each.

Christ will silence those who appeal to Mosaic Law to justify divorce by citing the biblical remark from this chapter: *"They will be two in one flesh"* and explicitly reaching the consequence of the indissolubility of marriage: *"What God has joined, let no man separate"*. It is only in Christ's teaching that the sacred dignity of cooperating with God to give life is maintained, providing a suitable environment for those who will also be "children of God" by being children of human love in a total and mutual self-giving.

Science has nothing to add directly to these theological remarks, but it does accept as objectively true that the human species is one, that all its members have an equal and rational nature, that they are also equally the subjects of rights and duties, and that Man is naturally meant to live in a society. The family is the basic unit of society and of every political structure, and, in the words of Pope John Paul II before the United Nations, "society is for the person, not the other way around". Only each human being is an "image and likeness" of God, called to an eternal destiny that concerns his personality above political or economic concerns of any kind. It is never acceptable to downgrade a human being to the level of "something useful" for scientific or economic progress, or to subject a person to any deprivation of the human rights that have their origin in the Creator, not in any artificial source of power, be it tyrannical or democratic.

Summing up

These first pages of the Bible are not outdated, and an intelligent and well informed believer has no reason to hide them as something childish and unsuited for this century.

Their teachings are as important today as they were thousands of years ago in the pre-scientific environment of the people for whom they were written.

It would be a mistake to read Genesis as if it were a book of Cosmology or Biology, thus being forced by a blind faith to deny scientific facts regarding cosmic evolution or the evolution of life on Earth. But, in a simple language, very important truths are presented, not to be forced artificially into a scientific straightjacket nor –as an alternative- dismissed by a myopic scientificism. It is possible to accept that each way of partially knowing the total reality of the Universe and Man will contribute to a richer synthesis that will benefit both scientists and believers. The center will be the Love of a Creator who is not satisfied with watching suns burn or with the scurrying activities of animals that cannot know Him.

A personal Creator –intelligent and free- can only have a sufficient reason to create in the desire to have personal relationships with beings similar to Himself and capable of partaking of his happiness outside of any time limit, because true love is forever. This is our destiny according to biblical Theology: to live, beyond all boundaries of time and space, in union with *"the Father from whom all good things come"*

Anthropic considerations
St. Peter Chrysologus (380 – 450) – Sermo 148: PL. 52, 596

"Why do you ask *how* you were created and do not seek to know *why* you were made? Was not this entire visible Universe made for your dwelling? It was for you that the light dispelled the overshadowing gloom; for your sake was the night regulated and the day measured, and for you were the heavens embellished with the varying brilliance of the Sun, the Moon and the stars. The Earth was adorned with flowers, groves and fruit; and the constant marvellous variety of lovely living things was created in the air, the fields, and the sea for you"...

"He has made you in his image that you might in your person make the invisible Creator present on Earth; He has made you his legate, so that the vast empire of the world might have the Lord's representative. Then in his mercy God assumed what He made in you; He wanted now to be truly manifest in Man, just as He had wished to be revealed in Man as in an image. Now He would be in reality what He had submitted to be in symbol"

St Alphonsus Liguori (1696-1787) – Sermon

"All the gifts which he bestowed on man were given to this end (*to gain our love*). He gave him a soul, made in his likeness, and endowed with memory, intellect and will; he gave him a body equipped with the senses; it was for him that he created heaven and earth and such an abundance of things. He made all these things out of love for man, so that all creation might serve man, and man in turn might love God out of gratitude for so many gifts."

"But he did not wish to give us only beautiful creatures; the truth is that to win for himself our love, he went so far as to bestow upon us the fullness of himself. The eternal Father went so far as to give us his only Son...he sent his beloved Son to make reparation for us and to call us back to a sinless life"..."By giving us his Son... he bestowed on us at once every good: grace, love and heaven; for all these goods are certainly inferior to the Son: *He who did not spare his own Son, but handed him over for all of us, how could he fail to give us along with his Son all good things?*"

Epilogue

Eighty years ago G. K. Chesterton (*St. Thomas Aquinas*, pp. 215-216, New York: Sheed and Ward, 1933) said:

> *For those who really think there is always something really unthinkable about the whole evolutionary cosmos as they conceive it; because it is something coming out of nothing; an ever increasing flood of water pouring out of an empty jug... In a word, the world does not explain itself and cannot do so merely by continuing to expand itself. But anyhow, it is absurd for the Evolutionist to complain that it is unthinkable for an admittedly unthinkable God too make everything out of nothing and the pretend that it is more thinkable that nothing should turn itself into everything.*

In other words, it is much more reasonable to think about the universe as **finite, open** and **contingent** than to think it as infinite and, at the same time, **coming out from nothing**. (See for instance M. M. Carreira, SJ, and Julio A. Gonzalo, *Everything coming out of nothing vs. a finite, open and contingent universe*, World Scientific: Bentham eBooks, 2012).

ABOUT THE AUTHORS

Father Manuel M. Carreira, SJ, after his degrees in Philosophy and Theology, he obtained a Master in Physics (John Carroll University, Cleveland) and PhD in Physics(The Catholic University of America, Washington) with a thesis on cosmic rays directed by Dr. Clyde Cowan (co-discoverer of the neutrino with F. Reines, Nobel 1955)

He taught Philosophy of Nature at Comillas University in Madrid. In 1991, on the centenary of the Vatican Observatory, he gave a course in Rome on galaxies for 20 bishops from different countries. In 1986, John Carroll University awarded him its Centennial Commemorative Medal in recognition of his work at that center. In 1999, he received the Medalla Castelao of the Galician regional government, for promoting the prestige to the region with his cultural activities

His book "Filosofía de la Naturaleza, Metafísica de la Materia", is used as a text in several educational centers in Spain and America. Two monographs, "El Creyente ante la Ciencia" (BAC) y "El Hombre en el Cosmos" (Sal Terrae) have been published in Spain.

Julio A. Gonzalo has a Ph. D. in Physics by the Universidad Complutense, Madrid. He did research and teaching at the universities of Salamanca (Spain), Mayagüez (Puerto Rico), Rio Piedras (Puerto Rico), Barcelona (Spain) and UAM Madrid (Spain). He worked as Research Collaborator for the US Atomic Energy Commission at Brookhaven National Laboratory, Upton, L.I., New York (1963-64) and as Division Head and Sr. Scientist at the PRNC-USAEC (1965-76). Now he is Professor Emeritus at the Universidad San Pablo- CEU, Madrid. He has been author or co-author of over 200 publications and a dozen scientific books, the latest "Cosmic Paradoxes" (World Scientific: Singapore, 2012).

www.ingramcontent.com/pod-product-compliance
Lightning Source LLC
Chambersburg PA
CBHW062005200326
41519CB00017B/4677